FINLAND
THE COUNTRY OF EVERGREEN FOREST

This book has been published in connection with the XXth IUFRO World Conference,
held in Tampere, Finland during 8.-12.8.1995

Layout: Kersti Tormis
Translation: John Derome
Colourseparations and printing: Forssan Kirjapaino Oy – Forssa Printing House Ltd
Paper: This publication is printed on Galerie Art Silk 130 g/m², which has been granted the Nordic Environmental Label.

ISBN 951-9026-33-9 Pohjoisen havumetsän maa
ISBN 951-9026-34-7 Finland the country over evergreen forest
ISBN 951-9026-35-5 Finnland ein Land der borealen Nadelwälder

SIMO HANNELIUS KULLERVO KUUSELA

Finland the country of evergreen forest

FORSSAN KIRJAPAINO OY

CONTENTS

PREFACE

This book has been written at the instigation of the organising committee of the IUFRO XX World Congress. The purpose of the book has been to present an overview of Finland's forests and their importance for the country's economy, and the principles of silviculture based on the ecology of the boreal coniferous forests. This perspective is important when striving to satisfy man's needs without endangering the sustainability of forest ecosystems or reducing their biodiversity in integrating Europe, as well as on a global scale. In order to achieve these goals we must understand the ecology of the forest zones and the silvicultural practices suitable for each zone which depend on it.

The Finnish Forest Research Institute and the European Forest Institute have provided the facilities required for this work. The Finnish Cultural Fund, the Timberjack Group and Kodak Finland have supported this undertaking. This support has especiallymade it possible to photograph the illustrations for the book during three summers.

Professors Eino Mälkönen, Jari Parvianen and Risto Seppälä, Licentiate Heikki Lindroos have made valuable comments on the manuscript. The expertise of innumerable researchers and foresters has been exploited in locating suitable photographic sites and obtaining background information about them. John Derome has translated the Finnish text into English.

We would like to thank all those who have supported, promoted and improved the manuscript.

The general section of the book has been prepared by Kullervo Kuusela, and the photographs and associated information by Simo Hannelius.

Helsinki, the 8th of January, 1995

Simo Hannelius *Kullervo Kuusela*

INTRODUCTION

When man first arrived in Finland conifer-dominated forests, interspersed by a mosaic of burnt areas, wetlands and watercourses, covered the land. The trees provided man with stored solar energy and raw-materials for dwellings, furniture, tools, carts, sledges, boats and ships. Wood ash fertilized the slash-and-burn fields and the forests provided extensive pastures. Charcoal and tar were made from wood.

The forest was a shelter against the wind and cold. It provided refuge during times of persecution. The old gods lived in the forest, it held the holy trees and bowers. The forest was also a threat. The wild beasts of the forest killed domestic animals and sometimes even harried man himself. The trees rapidly colonised any field or pasture left untended.

The people living in the forest were at the mercy of the harsh, cold climate and years of famine. Hunger, pestilence and war kept the population small. When their need was greatest, bread made from pine bast kept the toughest alive.

Man was ethically responsible for himself, his family and his tribe. For him the forests represented an endless supply of natural resources. He carved out fields and pasture from the forest. The forest met his needs and continued to renew itself.

Furs, wood and tar were the first export commodities. They were exchanged for goods unobtainable in Finland. As agricultural techniques evolved and developed, food production gradually shifted to ploughed fields and pastures. The need to invade and destroy the forest decreased. At the same time wood became the raw-material fuelling the fledgling industry producing goods for the domestic and export markets. Finland had discovered its green gold.

As the forest industries grew and the domestic demand for wood continued unabated, ever-increasing concern was voiced about the sufficiency of the forest resources. Rational methods of timber harvesting and silviculture were studied, developed and adopted. A sound basis was created for increasing prosperity and the development of Finnish culture. Man's responsibility expanded to cover the nation and its future.

As the living standard of the industrialised countries increased, the expanding world population became divided into developed rich countries and underdeveloped poor countries. The demands of industry started to exceed the sufficiency of local natural resources. The emission of pollutants by industry, traffic and energy production became a threat to the stability of the forest and other ecosystems, and started to undermine the very foundations of mankind's well-being. At the present time we are starting to realise that man's responsibility for satisfying his material needs must be expanded to cover the sustainability of ecosystem functioning and the biological diversity formed by the flora and fauna.

Man has to utilise natural resources to be able to live. However, he must do this without exhausting the natural resources, thus ensuring that enough remains for present and future generations. Plants and animals are values as such. The existence of no species should be jeopardised for the sake of short-term benefits.

Ecological stability and biodiversity presuppose that the utilisation and treatment of the self-renewing boreal coniferous forests are based on those laws of fire ecology which maintained the biological production and diversity of the forest before man first appeared. Acquiring sufficient information to achieve this goal is the joint task of national and international forestry research.

FINLAND'S POSITION IN THE BOREAL CONIFEROUS FOREST ZONE AND IN WORLD FORESTRY

The boreal coniferous forest zone forms an evergreen crown encircling the globe. It is bounded in the north by stunted forest and treeless tundra, and in the south by the temperate, mixed forest zone. This forms an intermediate zone between the coniferous forest and the deciduous, broad-leaved forest. The area of the boreal coniferous forest zone is about 900 million ha, which is 30% of the world's forests and 75% of the world's coniferous forests. The countries lying in the boreal coniferous forest zone are Alaska (USA) and Canada in North America, the Nordic countries Iceland, Norway, Sweden and Finland in Europe, and Russia. The major administrative areas of Russia are the Europe-Ural region, and Western Siberia, Eastern Siberia and the Far East in Asia.

The ice ages have forced the plants and animals to move south and again to the north in recurrent cycles. The number of tree species which form forests in Europe is therefore low, especially in those regions where major watercourses and mountain ranges run at right angles to the direction in which the species have had to move. In the Nordic countries the number of pioneer forest-forming, broad-leaved species is four, and of coniferous species two. Of the latter, Norway spruce is the only clear climax species.

The number of broad-leaved species forming forest in North America is 10, and of coniferous species 11. The corresponding numbers in Russia are 12 and 13. In addition to the Scots pine and Norway spruce of Fennoscandia, European Russia has some Siberian larch and Siberian fir. The high number of tree species in Asiatic Russia is partly due to the fact that most of the region was not covered by the last continental ice sheet. The number of tree species is highest in the Far East, especially in its SE part.

The forest fires that regulate tree species dynamics in natural conditions have nowadays been almost completely prevented in Fennoscandia. In Canada, on the other hand, they annually destroy 1 million hectares, and in Russia 1.5 - 2.0 million hectares. A special feature of the central and eastern parts of Canada is the regeneration cycle in spruce and fir forests which is maintained by insects that eat the needles and buds, thus killing the trees.

Regional differences in climate affect the fire ecology dynamics of the boreal coniferous zone. The summers in Canada and the central parts of Asiatic Russia are hot and precipitation relatively low. Forest fires recur frequently and they affect large areas. In regions with a maritime climate and high precipitation, such as in eastern Canada, the summer can be hot and dry for prolonged periods and fires destroy extensive areas of forest.

According to the accompanying cartogrammes, the forest area in Alaska is boreal coniferous forest proper. Apart from small-scale cuttings for firewood and building materials, it does not belong to the commercial forestry area. Alaska includes a narrow strip of coniferous rain forest along the Pacific coast.

There is also coniferous rain forest on the Canadian West Coast, and mixed forest and deciduous broad-leaved forest in the southern parts of the country.

Wood utilisation in Iceland's early history resulted in the natural forest becoming converted into birch scrub. An afforestation program is now being implemented using a number of exotic coniferous species. The area so far afforested is 10,000 ha.

In Russia the forest statistics cover the areas under the control of the governmental administration. 43 million ha of forest are in use by the state and co-operative farms.

Finland lies in the Eurasian taiga zone which stretches from the Atlantic to the Pacific. The flora and fauna of this great vegetation belt are more homogeneous than those of the other forest zones, and there are no local endemic species at all. The photograph was taken in northern Häme, central Finland.

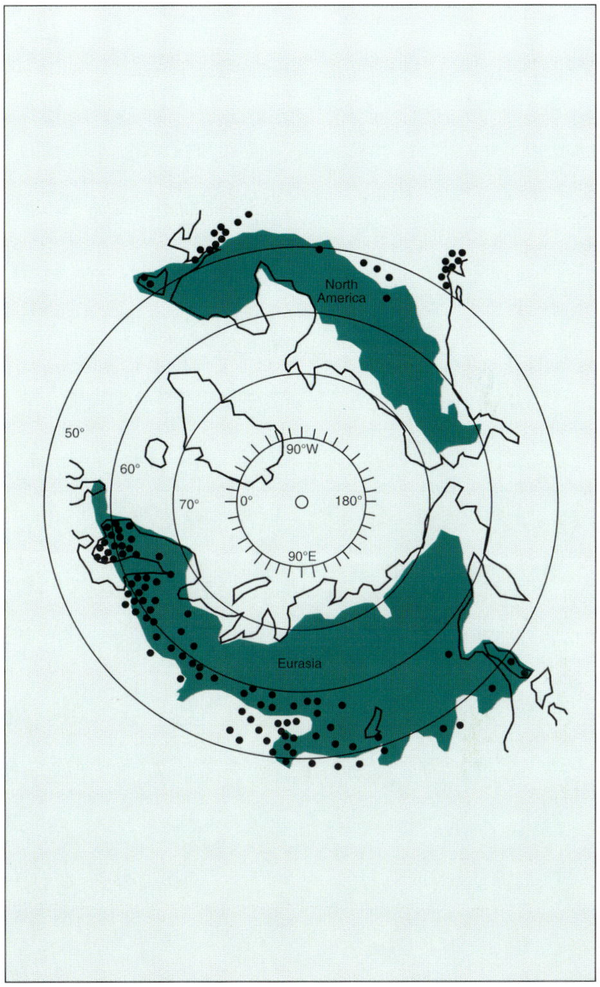

The northern coniferous forest forms an evergreen crown, extending around the globe between the treeless tundra and deciduous broad-leaved forest zones in North America and Eurasia. The points represent inhabited areas with a population of more than one million (1950). The population centres were concentrated along the southern edge of the coniferous zone, apart from the more evenly distributed centres in the Nordic countries and NW Russia.
Sources: Larsen, 1989, Oxford Atlas.

The concepts used in international forest statistics are not uniform. In order to improve their compatibility, an attempt has been made to convert estimates of the growing stock and its growth into over-bark stemwood volume.

In Fennoscandia the forest area refers to land where the potential production of overbark stemwood is at least 1 m^3/ha/year. In other countries the minimum yield limit is higher.

The net growth in the stem volume of the growing stock is the gross increment subtracted by the mortality. The Russian growth figures are calculated by dividing the standwise volume of the growing stock by stand age. The subsequent increase in the estimated growing stock is approximately the same as the net increment under conditions where there are many old and over-aged stands, and where thinnings are not carried out or their proportion out of the removals is very low.

The removals unit is the volume of round timber without bark that is harvested from the forest. The removals estimates presented here represent the production level in different countries at the end of the 1980's. The statistics for Russia refer to the situation at the beginning of the 1980's before the changes in the economic system started to have an effect.

The relative distribution of the population, forested area, growing stock volume and its net increment and removals between countries and sub-regions; the total area, forest area, growing stock volume and its net increment and removals per capita; and the proportion of conifer-dominated forest out of the total forest area and the growing stock volume and its net increment and removals per ha are presented in the maps on pages 13 and 14.

The large total area and high forest area per capita in Alaska, Canada and Asiatic Russia show that forestry in these regions primarily takes place outside the populated areas. This is also the case in the northernmost parts of European Russia. The costs of building roads and servicing work sites are much higher in these regions than in Fennoscandia, where land settlement extends rather deeply into the forests.

The proportion of exploitable forest out of the total forest area is highest in Finland (97%). The corresponding figures in Sweden and Norway are 90 and 67%, owing to the low profitability of forestry in mountainous areas. In Canada and Russia, especially in Asiatic Russia, there is a lot of forest which has so far remained outside the sphere of commercial forestry. In the regions with a cold climate and permafrost, as well as in mountainous areas, the forests remain permanently outside commercial exploitation.

The proportion of conifer-dominated forest ranges from 64% in Canada to 79% in Russia and 89% in Fennoscandia. The low proportion in the statistics for Canada is due to the fact that the forests are divided into conifer-dominated, mixed and broadleaf-dominated. The proportion of broadleaf-dominated forest has been maintained and increased through the reforestation of cutting and burnt areas by pioneer broadleaf species. There is a lot of birch-dominated forest in the Norwegian mountains.

The volume of the growing stock per hectare varies from 90 - 115 m^3. The variation in mean volume in the mountains and permafrost region in Alaska and the Russian Far East is 60 - 70 m^3/ha.

The annual net increment per hectare in the growing stock is highest in Fennoscandia, 2.1 - 3.8 m^3, with an average of 3.4 m^3. In Canada it is 1.7 m^3, in European Russia 1.9 m^3, in Western Siberia 1.3 m^3, in Eastern Siberia 1.1

m³ and in the Far East 0.9 m³. The main reasons for the low net increment values are the high proportion of old and over-aged stands, major fire and insect damage and a low level of thinning cuttings.

The annual harvest of bark-free timber per hectare is highest in Fennoscandia, 1.2 - 2.3 m³, and on the average 2.1 m³. In Canada it is 0.7 m³, in European Russia 1.4 m³ and in Asia 0.3 - 0.1 m³. Cuttings in the northern parts of European Russia were estimated at the beginning of the 1980's to be larger than the sustainable removals. The major problem there is the shifting of cuttings further and further away from the sites of consumption, into areas where the timber quality is poorer.

The net increment and removals per hectare statistics show that silvicultural practices, thinnings and the rapid establishment of new seedling stands after final cutting, can increase the sustainable timber production per unit area by a factor of two or three compared to the conditions in regions where intensive forestry is practised.

The higher the proportion of forest resources per capita, the greater is the role played by forestry and the forest industries in the national economy. The annual removals in the Nordic countries is 2.6 - 9.2 m³, in Canada 6.7 m³ and in Russia 2.6 m³ per capita, while the world average is 0.7 m³. The coniferous forests account for about 19% of the total removals in the world, 33% of the industrial wood and 35% of the industrial softwood cuttings.

		North America	Nordic countries	Russia
Broadleaves				
Grey alder	*Alnus sp.*	2	1	2
Aspen and poplar	*Populus sp.*	4	1	3
Birches	*Betula sp.*	4	2	7
Pines	*Pinus sp.*	4	1	3
Larches	*Larix sp.*	1	–	2
Shade trees				
Spruces	*Picea sp.*	3	1	4
Firs	*Abies sp.*	1	–	4
Thujas	*Thuija sp.*	1	–	–
Hemlocks	*Tsuga sp.*	1	–	–
Total		21	6	25

The number of forest-forming tree species in the boreal coniferous forest zone.

Relative distribution of the population and forest resources in the countries of the boreal coniferous forest zone.

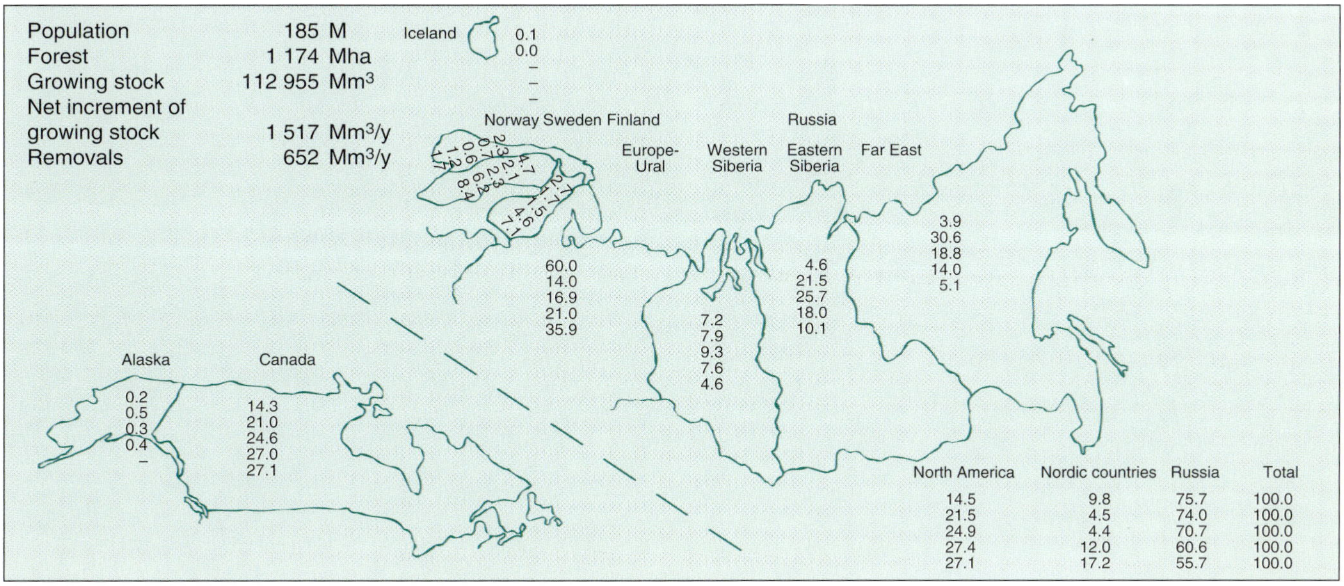

Surface area and forest resources per capita in the countries of the boreal coniferous forest zone.

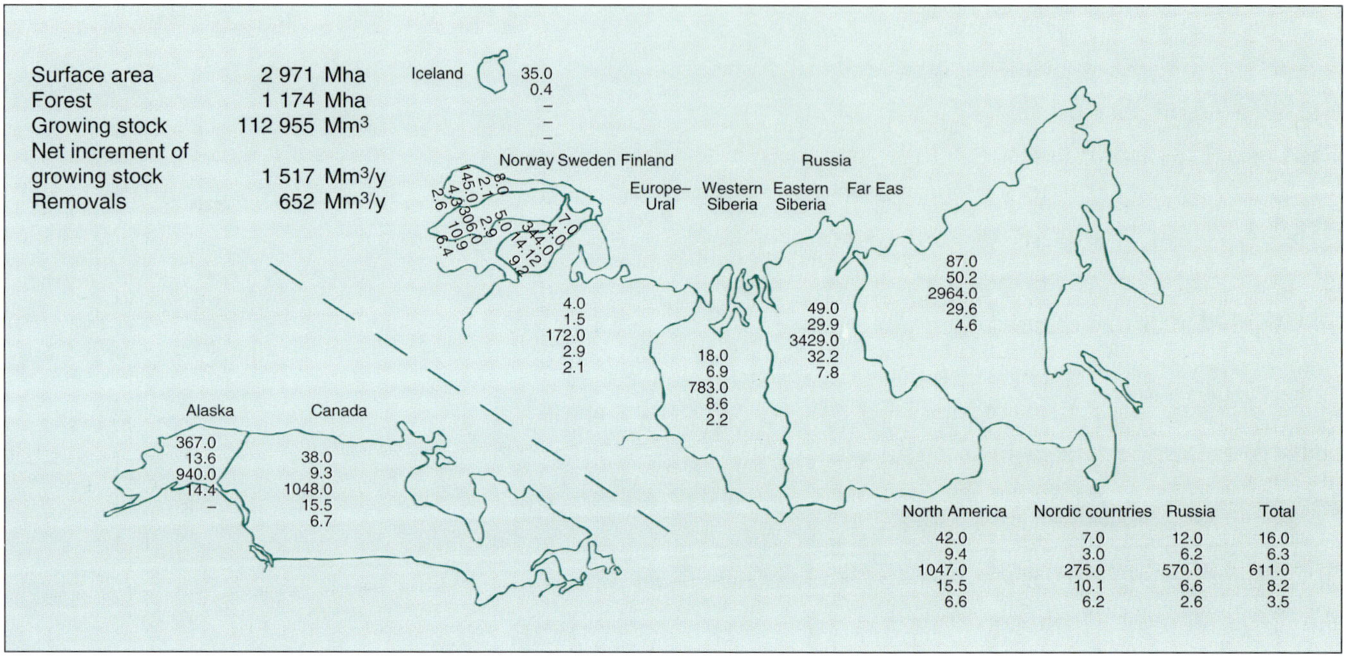

Proportion (in %) of conifer-dominated forest out of the total forest area, and the growing stock, net increment and removals per hectare.

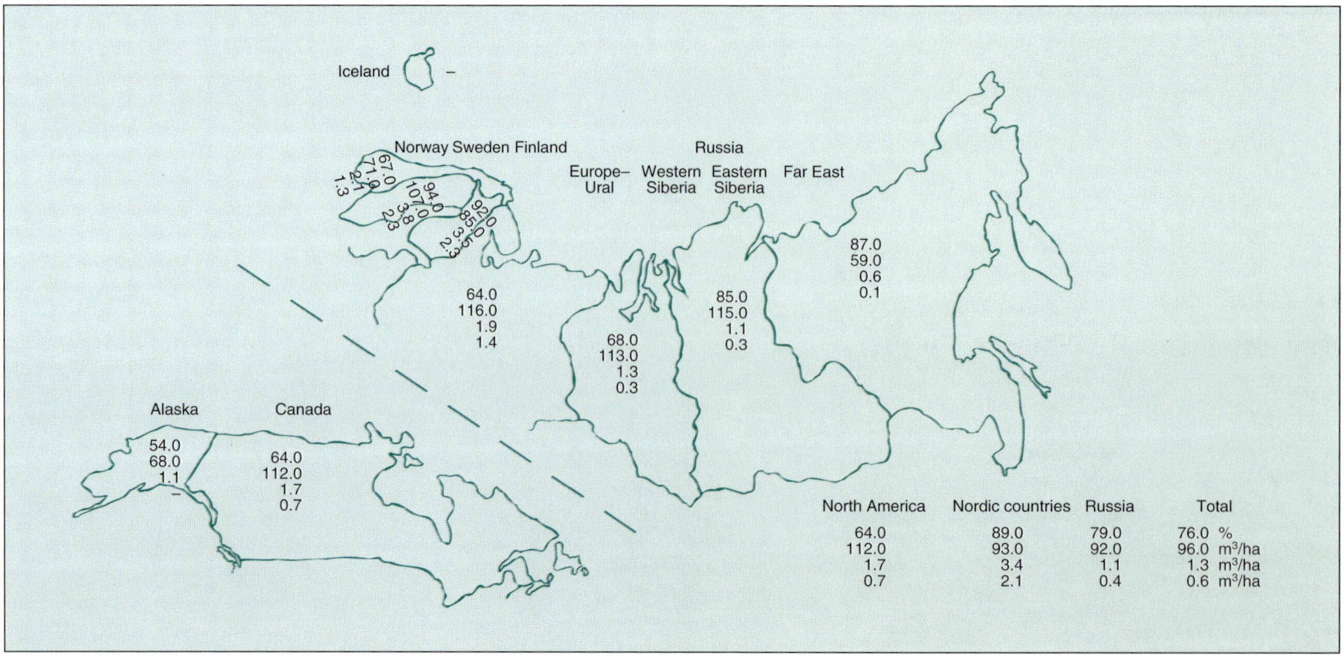

THE ECOLOGICAL PRINCIPLES OF SILVICULTURE

A profound understanding that boreal coniferous forests are ecosystems governed by natural laws has been the foundation of Finnish silviculture and forest management. In practical forestry, the fertility of the soil is estimated using a system of forest site types defined on the basis of the ground vegetation. The individual tree species are grown on the forest site types natural for each tree. Despite a considerable amount of research and field trials, the tree species used in commercial forests in Finland are still domestic ones. Tree growth, self-thinning and regeneration in natural forests are imitated in the commercial growing and regeneration of the stands.

The conflict between the growing needs of an expanding world population and the finite nature of natural resources, the sustainable development strategies based on international agreements, and the increasing amount of information about the functioning of ecosystems and their biodiversity, all presuppose the development of forest management in such a way that meeting the diverse needs of mankind and preserving the biodiversity of the forests are integrated into a balanced whole. The principles of forest ecology form a basis for understanding traditional and present-day forestry and for directing the development of forest management.

THE BOREAL CONIFEROUS FOREST ZONE AND THE PARTS OF THE FOREST

Forests are the most complex communities of animals and plants on the earth's surface. The forests of the world are divided into forest zones, the species composition of which is dependent on the climate and soil. They are the boreal coniferous forest zone, the temperate deciduous broad-leaved forest zone, the warm and dry evergreen forest zone, the dry and hot savannah forest zone and the tropical rain forest zone. The boreal coniferous forest zone is bounded in the north by the forest tundra and treeless tundra. There are transition belts between the forest zones proper. The southern edge of the boreal coniferous forest zone is bordered by mixed forest, which is characterised by a gradual transition from coniferous forest into deciduous broad-leaved forest. Conifers are the dominating tree species in coniferous forests, and on moving to the south the competitivity of broad-leaved trees gradually improves over that of the conifers.

Almost all of Finland is located in the boreal coniferous forest zone (taiga), which is divided into a northern, central and southern sub-zone. The mixed forest zone occupies a narrow strip along the SW coast and in the archipelago.

The forest consists of a collection of individual stands in which the soil quality, tree species composition and tree age are approximately uniform, and different from those in the surrounding stands. The tree stand is formed from tree species and individual trees.

Distinguishing between the individual components of the forest is facilitated by an understanding of tree stand dynamics. The death of individual trees is not necessarily a sign of forest death if it is a question of the natural, internal dynamics of the stand. When trees grow and compete for growing space, some of the trees lose their vitality and die. A weaker species may give way to a stronger species on the same site. Self thinning of the tree stand is a normal phenomenon in the dynamics of a healthy forest. The forest only dies when a change in the environmental conditions converts it into a treeless plant community or plantless desert.

The growth factors deteriorate and tree growth decreases on moving northwards or with increasing altitude in the mountains. The growth factors are the most unfavourable for trees close to the timber line, where the closed forest ends and the vegetation gradually changes into isolated groups of stunted trees, and finally into treeless plant communities.

Forestry land is divided for timber production purposes into three classes according to how well it supports tree growth. On forest land the growth factors are so favourable that the tree trunks are straight and tall and the tree crowns close to form a fully stocked stand. On scrub forest land the trees are short, branchy and often have a bushy form. Waste land lacks or almost lacks a tree cover. Multiple forest use is possible on all types of forestry land. Wild berries and edible mushrooms grow on scrub and waste forest land. Reindeers also graze on them.

The vertical distribution of the plants growing in a forest stand is divided into distinct vegetation layers. The trees form the uppermost layer, with the shrub layer below it. The ground vegetation proper is divided into tall dwarf shrubs, grasses and herbs in the upper ground vegetation layer, and short dwarf shrubs, mosses and lichens in the lower layer.

Silviculture and forest management is primarily concerned with the dominant tree storey. This storey may have older hold-over trees lying above it, and a younger tree understorey below it.

In the dominant storey a distinction is drawn between the dominant and the dominated crown layers, the former layer comprising the trees which are the main goal of commercial forestry. The dominant crown layer is divided into main dominant and co-dominant trees, and the dominated layer into intermediate and suppressed trees.

THE FUNCTIONAL COMPONENTS OF FOREST ECOSYSTEMS

Solar energy, climate, soil and associated physical and chemical properties form the framework for the living components of forest ecosystems: plants, animals and micro-organisms. Examination of the functioning of ecosystems and the silvicultural methods which take these aspects into account is made easier by dividing the whole ecosystem into its functional components. These are the throughflow of solar energy, climate, the water cycle, bedrock and soil types, the living soil and soil formation, trees as assimilating plants, fauna in the nutrient web of the forest ecosystem, and nutrient cycling.

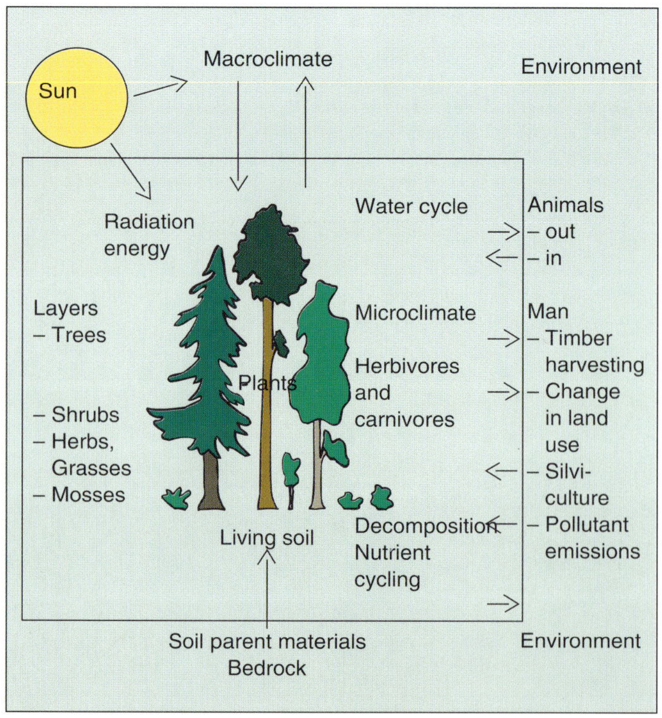

The forest ecosystem and its individual components, and the environmental factors affecting the ecosystem.

Scots pine (Pinus sylvestris) forms the timber line in the Nordic countries; in the Kola Peninsula, NW Russia, it also includes Norway spruce (Picea abies). Pine competes with mountain birch (Betula tortuosa), which is considered to be a subspecies of pubescent birch (Betula pubescens). The pine timber line is moving to the north owing to climatic warming during the 1930's. In 1965 the autumnal moth (Epirrita atumnata) caused extensive damage to mountain birch. This promoted pine regeneration between Inari and Karigasniemi in northern Lapland. The photograph was taken in 1993.

One of Finland's northernmost, artificially regenerated stands is located at Näätämö within the closed timber line to the north of Lake Inari. The effective temperature sum in the area is about 600 d.d. and the altitude 90 m a.s.l. The 350 ha of plantations were established during 1921-1939 primarily by sowing after a forest fire. The seed originate from near the Arctic Circle, about 350 km to the south. The density of the growing stock is 1 200 stems/ha, the increment ranging from 1 to 2 m³/ha/year with a height growth of 15 cm/year.

A pine forest that has developed after prescribed burning by Prof. Gustav Sirén in 1952 along the timber line at Muotkanruoktu, halfway between Inari and Karigasniemi in northern Lapland. The seedling stand has regenerated naturally. Movement of the timber line towards the north is regulated by seed ripening. During the post-glacial warm period about 5 000 years ago the timber line was located 200 m higher up the fells.

Temperature is the minimum factor determining seed ripening. Large rocks on southern slopes improve the microclimate close to the ground surface. This promotes seed ripening, stabilisation of the timber line and the spread of the trees to the north as the climate warms up.

Spruce grows together with mountain birch in mixed stands; the spruce timber line is located on the Saariselkä and Pallas fells south of Lake Inari. Spruce occurs singly and in small groups to the north of the timber line. Under extreme climatic conditions spruce regenerates vegetatively by means of roots formed on branches in contact with the ground. The photograph was taken on the Saariselkä fells at an altitude of 350 m a.s.l.

Erosion is a rare phenomenon in the boreal coniferous forest zone. Fine-textured soils deposited by wind action are sensitive to erosion if forest fire destroys the binding layer of vegetation and humus. Pyrrinkangas (near Tyrnävä) is located close to the Ostrobothnian coast at an altitude of 40 m a.s.l. Over 1 000 ha of forest were destroyed in a fire started by lightning in 1969. Although pine seed has been repeatedly sown in the area, reforestation had not been completely successful by 1992.

The border between the boreal coniferous forest and the temperate broad-leaved forest zones passes along the southern coast of Finland. The only area in Finland where beech (Fagus sylvatica) forms closed stands is the Åland Islands.

The throughflow of solar energy

The more direct the solar radiation falling on the earth's surface, the greater the amount of solar energy available for the vegetation. The incoming angle and amount of solar radiation per unit area decrease on moving towards the polar regions. Finland's position between latitudes 60°N and 70°N means that the climate is relatively unfavourable for plant growth. However, the long day and abundant solar radiation during the growing season compensate to some extent for this unfavourable position.

More solar energy falls on southern-facing slopes than on northern ones. In Finland's undulating, but rather flat terrain this factor is only of any real importance close to the northern timber line where the trees grow best on southern and south-western slopes.

Cloudiness reduces the amount of solar energy reaching the ground. The average frequency of cloudy days is greater in the autumn when the trees' growing season has terminated. During exceptionally cloudy growing periods the ground is cooler than average, which slows down the decomposition of plant litter and retards nutrient cycling.

Part of the total radiation is reflected back from the surface of the vegetation and ground into the atmosphere. The net energy retained at the ground surface is relatively constant during the growing season in areas with a closed forest cover. It is smaller in the forest tundra and tundra zones. The amount of outgoing energy is greatest during the winter and on treeless land. During the permanent snow cover period it is low in forest with a closed canopy.

The density of the dominating tree canopy effectively regulates the distribution of solar radiation between the ground and the different vegetation layers. In a fully stocked stand the ground vegetation and soil each receive only 2% of the total incoming radiation. For this reason the soil temperature in a dense stand is relatively low.

The energy needed to support the biological production and hydrological cycle of green plants is obtained from solar radiation. Only about 2% of the total radiation energy is bound in the gross biological production of trees. Half of this is used in respiration and the other half is bound in the biomass as net production. The energy bound in the biomass is transferred to the other components of the ecosystem's nutrient web and, through them, is gradually lost back into outer space.

Climate

Climate can be divided into two components: the macroclimate and the microclimate. The macroclimate is independent of the degree of forest cover and type of forest management. The most important macroclimate parameter from the point of view of tree growth is air temperature. Other important parameters, which are in fact dependent on air temperature, are the length of the growing season and the effective temperature sum, amount of precipitation, windiness and degree of cloudiness. The deposition of inorganic material from the atmosphere also has an effect on tree growth. Greater amounts are deposited in forested than in open areas.

The annual mean temperature on the south-western coast of Finland is +5°C. It gradually decreases on moving to the north-east and north, and is about -1°C at the timber line. Owing to the warm, moist air masses derived from the Gulf Stream and Atlantic, the climate in NW Europe is warmer than that in other areas at the same latitude in other regions of the world. The Baltic Sea and

Arctic Ocean dampen the temperature extremes in adjacent areas.

The mean temperature during the warmest month (July) is +17.5°C on the south-western coast and +13°C at the timber line. During the coldest month (February) the corresponding temperatures are -5°C and -13°C. The mean temperature in February in the NE corner of Lapland is -10°C owing to the proximity of the Arctic Ocean. The temperature during the summer can rise to above +30°C, and during the winter to below -40°C.

The length of the thermal growing season is the number of days when the mean daily temperature is at least +5°C. The length of the growing season in the region with a closed forest cover ranges from 120 to 180 days. The effective temperature sum of the growing season, expressed as degree days (d.d.), is the sum of the number of degrees (above the threshold value of +5°C) on those days during the growing season when the temperature exceeds +5°C. The effective temperature sum ranges from 700 to 1300 d.d.

The length of the growing season, and the effective temperature sum especially, are the climatic variables which best explain tree growth, flowering and the proportion of germinating seed.

The local climate, which depends on the terrain and exposition of the slopes, can differ considerably from the macroclimate. Thus a warmer than average local climate can, especially close to the timber line, result in the formation of germinating seed in areas where the macroclimate is too cold for seed maturation.

Long-term mean values do not provide any information about the large variation in the annual temperature conditions. The effect of the coldest and warmest summers during this century on tree growth has been the same irrespective of whether the growing site would be moved 500 km to the north or south in a region with the same mean macroclimate. Trees in the boreal coniferous forest zone are characterised by an ability to withstand extreme values in the annual, monthly and daily temperatures.

In areas with an exceptionally cold winter, frost and desiccating wind turn conifer needles brown and cause frost cracks in the trunks of broad-leaved trees. Severe frost can damage the fine roots when the snow cover is thin or completely absent. Summer frosts mainly damage the new shoots on spruce.

The mean annual precipitation is 650-700 mm on the southern coast, 500-550 mm on the western coast, 550 mm in southern Lapland and 400 mm at the timber line. The annual evaporation of 425 mm in the south and 200 mm at the timber line is clearly less than the annual precipitation. However, evaporation is greater than precipitation during July in the south-western parts of the country and in inland areas. It is less than precipitation at all times of the year in northern Finland.

Owing to the cool climate, evaporation does not normally result in a lack of water for plants apart from during drier and hotter than average summers in the southern archipelago, on rocky sites along the coast, and on pervious sand and gravel soils. The relationship between precipitation and evaporation results in the formation of waterlogged areas that gradually develop into peatlands .

About 30% of the precipitation falls as snow on the SW coast, over 30% in western Finland, almost 40% in eastern Finland and slightly more in Lapland. There is a permanent snow cover for 4 - 5 months in southern Finland, for 5 - 7 months in northern Finland and for even longer on the fells in the north. The thickness of the snowpack is 20 cm in the Åland Islands, 40-50 cm in southern Finland, 70 cm in the eastern and northern parts, 100-200 cm on eastern hilly sites in northern Finland, and 50-60 cm in the northernmost parts of Lapland.

The snow pack protects plants and animals at the ground surface from freezing temperatures. The ice and snow which builds up in the tree crowns break the branches and leaders of trees in the hilly areas in the eastern and northern parts of the country. Spruce, which best withstands snow damage, has a competitive advantage over other tree species in these regions.

Storm damage in Finland is of relatively little importance compared to other parts of the world. Under normal conditions the wind mainly fells trees with a rot-weakened trunk and roots, or which have a shallow rot system on waterlogged sites. Even the most serious cases of storm damage in Finland are minor compared to the situation in western and central Europe.

The atmosphere consists of 78% nitrogen, 21% oxygen and 0.03% carbon dioxide. However, the nitrogen is in a form which plants are not able to use. Although the proportion of carbon dioxide is small, it is an essential raw material for assimilation and it regulates the temperature of the lower layers of the atmosphere.

Natural atmospheric deposition falling on the ground contains plant nutrients such as nitrogen oxides and sulphur. However, the emissions from industry and traffic have changed the chemical composition of deposition to such an extent that it nowadays has an effect on the floral and faunal composition and biological production.

The climate within the forest is called the microclimate, and the extreme values are much smaller than those of the macroclimate. There is less light, more moisture and abundant carbon dioxide produced by the respiration of living organisms close to the ground surface than higher

During the Ice Age huge blocks of ice remained embedded in the soil in the esker areas. When they finally melted they left conical, steep-sided pits. The microclimate in the pits is unfavourable for plants because cold air (frost) collects at the bottom in early summer. The phenomenon occurs in southern and central Finland. The photograph was taken in Northern Karelia. The new spruce buds are very sensitive to frost damage that occurs when the temperature falls below -3 to -4°C. Spruce is the first tree to be damaged, and repeated frost kills the seedlings.

Frost and high humidity in late spring predispose pines growing in low-lying parts of the watershed district in southern Finland to attack by Sclerroderis canker (*Gremmeniella abietina*). Sclerroderis canker occurs in all types of stand, but young trees are the most seriously affected. Summer frost also nips the sensitive buds of pine leader shoots; damage occurs when the temperature falls below -5°C. The photograph was taken in northern Häme.

The crowns of hilltop spruce stands become covered with a thick layer of snow and condensed ice. The slopes with the thickest snow cover occur at altitudes of over 200 m. Spruce is able to carry the heavy load of ice and snow. The photograph was taken on Pisavaara close to the Arctic Circle.

During winters with heavy snowfall the load of ice and snow can be as much as 100...150 kg/m of stem. Spruce withstands the load, but pine and birch are easily broken on the slopes of Pisavaara. Snow thus promotes the development of pure spruce stands. Pure spruce stands regenerate slowly in the small openings.

Pine grows well and regenerates naturally in the valleys of Lake Inari. The seedtree method and light soil preparation are suitable silvicultural methods. The climate in the area is favourable owing to the proximity of the Gulf Stream and the temperature-moderating effect of the large lake basin.

up in the plant canopy and above it. The air temperature is lower in a fully stocked coniferous stand and the wind velocity lower than that outside the forest.

The hydrological cycle

Water is the one of the raw materials in plant assimilation. It is also the vector for the nutrients required by the plants. The water in forest soils is primarily derived from precipitation, although running surface water and the ground water can also play an important role. A raised bog is an example of an ecosystem whose only water supply is that derived from precipitation. The surface water containing nutrients and oxygen that flows down slopes increases tree growth in comparison to the situation on a flat site composed of the same type of soil.

The tree canopies in a fully stocked stand intercept a fairly high proportion of the water falling as rainfall. Some of the intercepted water evaporates, and the rest falls to the ground as canopy drip and stemflow. Part of the water taken up by plants is lost immediately as evapo-transpiration, and part is gradually liberated in the respiration of plants and animals.

Meteorological parameters in Finland

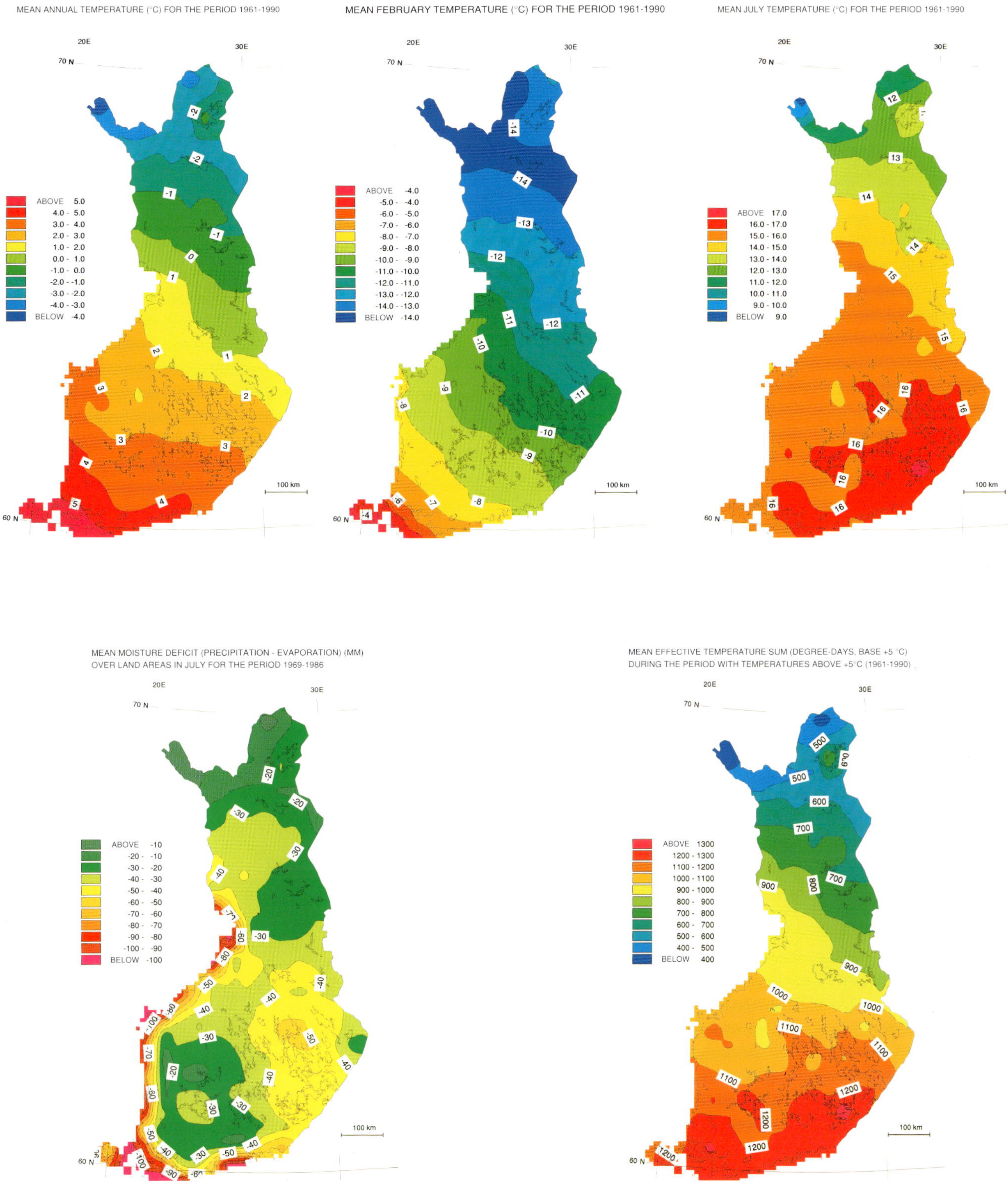

The forest is the greatest consumer and evaporator of water in natural plant communities. The forest regulates the flux of water and maintains the quality of the ground water. Deforestation increases flooding and causes erosion in areas where there is no ground vegetation, humus or peat binding the soil. The hydrological cycle is powered by solar radiation. Heat energy circulates along with the water in this cycle.

Bedrock and soil types

Granite and gneiss are the most common types of rock in Finland. The minerals they are composed of, feldspars, schists and quartz, are poor in nutrients and contain little basic material. Limestone and dolomite occur in small deposits in southern Finland. The Karelian schists in eastern and northern Finland usually contain more nutrients. The last continental ice sheet moved the overburden, i.e. the loose soil and stones, to the south and south-west. The overburden in most areas is thin, with the bedrock close to the surface or exposed.

The soils have been fashioned by the ice and melt water and transported to their present locations. Glacial tills are found close to their parent bedrocks. They contain particles of all sizes, ranging from clay to boulders. Soils which have been transported and sorted by water movement, i.e. sorted gravel, sand, fine sand and clay soils, consist of weathered and rounded particles of approximately the same size.

About 80% of the area of upland soils consists of glacial till. The particles and stones in this type of soil are angular and rough. The different particle size fractions are usually mixed together. Morainic areas are characterised by hills and hollows, and stony scree and exposed bedrock. Owing to the stoniness, it is difficult to convert them into fields, and most of them have remained as forest land. They contain fine sand and clay and the forest grows well on them.

The eskers formed from sorted soil account for 5% of the upland sites. Stones, gravel and sand often occur in separate, overlying strata. They usually have a low content of fine sandy material, and silt and clay are almost completely lacking. They are therefore usually rather infertile sites. The longitudinal eskers, formed by the water flowing off the glaciers, give a characteristic touch to the landscape in Lakeland Finland. The southern edge of the Salpausselkä esker area is the largest terminal ridge formation, and marks the point where the retreat of the ice sheet was interrupted at two stages, 11 200 - 10 950 and 10 450 - 10 250 years ago.

The proportion of sandy soils is 11%. They are composed almost exclusively of particles of the same size and contain few nutrients. They form extensive dry upland areas, especially in parts of Satakunta and along the coast of Ostrobothnia.

The proportion of silt and clay soils is 4%. They also consist of particles of approximately the same size, and when loosely compacted they make the best forest soils. Silt soils are usually compact and impervious to water, which reduces tree growth. Clay soils are to be found in the provinces of Uusimaa and Ostrobothnia and in south-west Finland, and to a lesser extent in inland regions which were covered by water after the last ice age. Most of these soils have been converted into fields.

The particle size is largest in gravel and coarse sandy soils. It is smaller in fine sand and in silt, and the smallest in clay. The greater the proportion of clay particles in the soil, the greater is the amount of nutrients such as potassium and calcium, and the better the water-retaining capacity.

The humus layer in upland soils and the peat organic soils are formed from the remains of plants and animals. Humus accumulates on top of the mineral soil and becomes partly mixed into the uppermost mineral soil layer. Plant remains, especially conifer needles, decompose slowly in the cool, moist climate, as well as on sites with mineral soil containing few nutrients, especially acidity-reducing calcium. Many of the nutrients in partially decomposed humus are in a form unavailable to plants, and the tree growth potential is lower. When the mineral soil contains a lot of nutrients, especially calcium, the humus consists of neutral or slightly acidic mull.

Peat is formed on wetlands where much of the plant remains are located below the water level. There is a lack of oxygen (anaerobic conditions) and decomposition is greatly retarded. The thickness of the peat layer varies from a few centimetres to as much as 10 meters.

Trees grow best on peatlands if their roots obtain nutrients from below the peat and from the water flowing in from the surrounding mineral soil. The trees are stunted or completely absent on thick-peated peatlands where all the nutrients are obtained from precipitation, or where the ground water table is close to the ground surface.

Two thirds of the forestry land, i.e. forest land, scrub land and waste land, is upland soil and one third peatland. 76% of the forest land is mineral soil and 24% peatland.

The area burnt by the major forest fire at Tuntsa in eastern Lapland consists of stratified, smooth-textured gravely soils where pine grows well. The few large pines that survived the fire in 1960 have produced even-aged, thin-branched young stands.

Fine-textured glacial till soils, with a high content of large boulders, are typical sites for spruce. After the Tuntsa fire they were first covered by a natural growth of birch, with a gradually developing understorey of spruce.

Eskers, deltas, ice-marginal and interlobate formations, and clay deposit

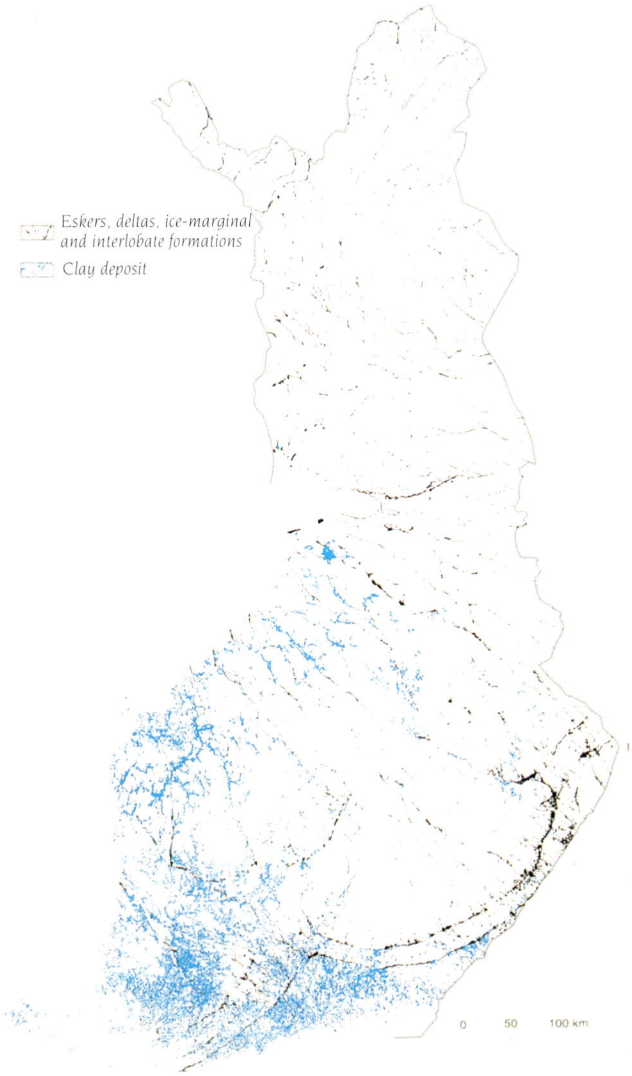

Provinces and topography in Finland

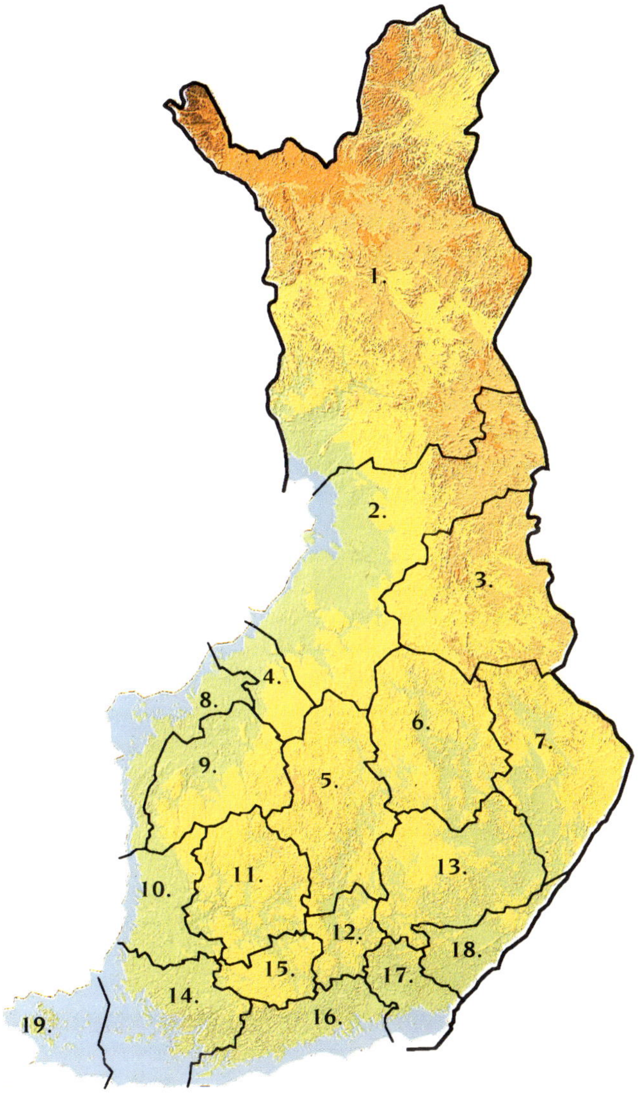

1. Lapland
2. Northern Ostrobothnia
3. Kainuu
4. Central Ostrobothnia
5. Central Finland
6. Northern Savo
7. Northern Karelia
8. Swedish Speaking Ostrobothnia
9. Southern Ostrobothnia
10. Satakunta
11. Northern Häme
12. Eastern Häme
13. Southern Savo
14. Southwest Finland
15. Southern Häme
16. Uusimaa
17. Kymenmaa
18. Southern Karelia
19. Åland Islands

The living soil and soil formation

The living soil consists of the loose mineral soil formed as a result of physical weathering, and humus. It contains roots, animals, fungi and micro-organisms. The soil solution contains chemically active clay and humus particles, on the surface of which the positively and negatively charged nutrient ions, utilisable by plants, are bound.

Natural rainwater contains small amounts of sulphuric and nitric acids which, when ionised in water, form positively charged protons and negatively charged sulphate and nitrate ions. The protons displace the nutrient cations from the surfaces of the clay and humus particles. If the plants do not have time to take up these ions, they are leached from the surface layers of the soil and soil fertility is reduced.

The concentration of protons in the soil, i.e. acidity, is an important parameter in the nutrient system. The higher the pH value (used to measure acidity), the lower is the acidity of the soil and the more nutrients it usually contains. The nutrient status of soils with a similar pH range is also dependent on the soil type, moisture conditions and composition of the vegetation.

Humus pH on fertile soils ranges from 5.0 to 5.5, on good till soils 4.7, and about 4 on poor sandy soils. The pH of the subsoil ranges from 5.3 - 5.8. The pH of the soil along the coast of Ostrobothnia, which is gradually rising up out of the sea, is about 7.

Soil formation is a process based on the weathering of the mineral soil, and the transport by water of the dissolved material from one layer to another. Precipitation is greater than evaporation in the boreal coniferous forest zone and the excess water percolates down through the soil. The transport of dissolved material from the topmost soil layers down into the underlying soil results in the formation of stratified soils called podsols.

The uppermost layer, the E or eluvial horizon, is of a greyish colour which becomes lighter with increasing depth. It is the most strongly weathered and leached of the different horizons. Relatively high acidity accelerates the leaching processes. Aluminium, iron and silicon are also released from the minerals in this horizon as a result of weathering, and are carried deeper down into the soil. Aluminium, iron, silicon and organic matter are precipitated in the underlying horizon (B or illuvial horizon). The upper part of this horizon (B1) is usually of a brownish colour owing to the precipitation of iron oxides, and in places black as a result of precipitated organic matter. The lower part of the B horizon is usually dark, reddish brown or light yellow.

Podsols are the most common type of soil on sandy soils which are highly permeable to water. The degree of leaching and podsolisation decreases with increasing clay content.

Trees as assimilating plants

Green plants utilise solar energy to produce carbohydrates from carbon dioxide and water. At the same time oxygen, water and heat energy are released. The gross biological production is the result of assimilation. The level of gross biological production is the greater, the higher the air temperature during the growing season and the more solar radiation, water, carbon dioxide and nutrients there are available. The dependence of assimilation on the air temperature in conifers and broad-leaved trees is illustrated in the following:

When precipitation is greater than evaporation, water percolates down through the soil. The compounds dissolved in the soil water precipitate after moving a certain distance down the soil profile. Podsolisation is the most common type of soil-forming process in Fennoscandia: the soil profile consists of overlying litter, humus, eluvial and illuvial layers. The soil below the illuvial layer is the pristine subsoil. The eluvial layer is usually light grey in colour, and the illuvial layer is a reddish-brown colour due to the precipitation of iron oxides. The photograph was taken in the Kuusamo area (eastern Finland), which was not covered by water after the last Ice Age.

Assimilation		Lower limit	Optimum	Upper limit
Conifers	°C	–5...–3	+10...+25	+30...+40
Broad-leaves	°C	–3...–1	+15...+25	+40...+45

Gross biological production commences at the lower limit of the temperature range, is greatest within the optimum temperature range, and ceases at the upper limit. These limit values clearly show that the competitivity of conifers is better than that of broad-leaved trees in the cool climate of the boreal coniferous zone.

Trees consume part of their gross production in respiration in order to maintain their metabolic functions. Respiration consumption starts at a temperature of about 0°C. As the temperature rises it increases at a slower rate than the gross production, reaching a maximum at about +55°C. Respiration consumption ceases at about +60°C.

The net production, which consists of carbohydrates, is the difference between gross production and respiration. Net production is at its greatest within the temperature range +18 to +27°C. Trees produce biomass from these carbohydrates and the mineral nutrients taken up from the soil.

Respiration consumption can be divided into the consumption of carbohydrates in maintaining the plant's metabolic functions, and the consumption involved in increasing the plant's biomass. Maintenance consumption in a stand increases with increasing age and biomass of the trees. The older the stand and the greater its biomass, the higher is the proportion of gross production consumed in maintenance respiration, and the smaller the proportion that is available for increasing biomass. When the stand reaches an age where death and decay of the trees starts to become appreciable, stand biomass also decreases.

Growth in stemwood volume is a part of net production. The development of volume growth with respect to stand age is approximately the same as the overall development of net production. Volume growth increases rapidly after formation of the seedling stand, reaches a peak when the stand is middle aged, and then gradually decreases. The age when thinning and final cutting are performed in commercial forests is based on the development of net growth in natural stands

The respiration consumption of trees continues at a low level at temperatures where assimilation has ceased. During mild winters when the temperature is around 0°C or slightly above, the trees consume the carbohydrates, produced during the previous summer, which were primarily intended for use during the coming summer. A colder than average preceding winter is more favourable than a mild one from the point of view of maximum net growth during the growing season. The height growth of conifers is based on about 60% of the carbohydrates stored during the previous summer, and about 40% of those produced during the current growing season.

High biomass and stem volume growth presupposes efficient assimilation. A high sugar content (carbohydrates) improves the ability of trees to resist fungal attack.

The foliage of conifers consists of 2 - 9 needle years, spruce having more than pine. The oldest needles are shed as the new needle age classes are produced. About 80% of assimilation takes place in the needles of the two youngest needle years. If the nutrient balance in the trees is disturbed, nutrients can be transferred from some of the oldest needles, and these needles then shed without any significant decrease in assimilation.

In conifers needles are available for assimilation immediately at the start of the growing season. Deciduous trees first have to produce their assimilating organs (i.e. leaves or needles) using the carbohydrates and mineral nutrients stored during the previous summer. This represents a considerable advantage for conifers in conditions where the growing season is short.

Tree species which tolerate shade have a lower metabolic rate than light-demanding species. Compared to light trees, the amount of carbon dioxide which they assimilate in low-light conditions is greater than the amount they produce in respiration. This improves the competitiveness of spruce compared to other tree species.

Schematic diagram of gross biological production and its distribution between tree respiration and net growth with increasing stand age. Gross stemwood increment, mortality and net growth for a natural Scots pine stand growing in southern Finland on a Vaccinium forest site type.

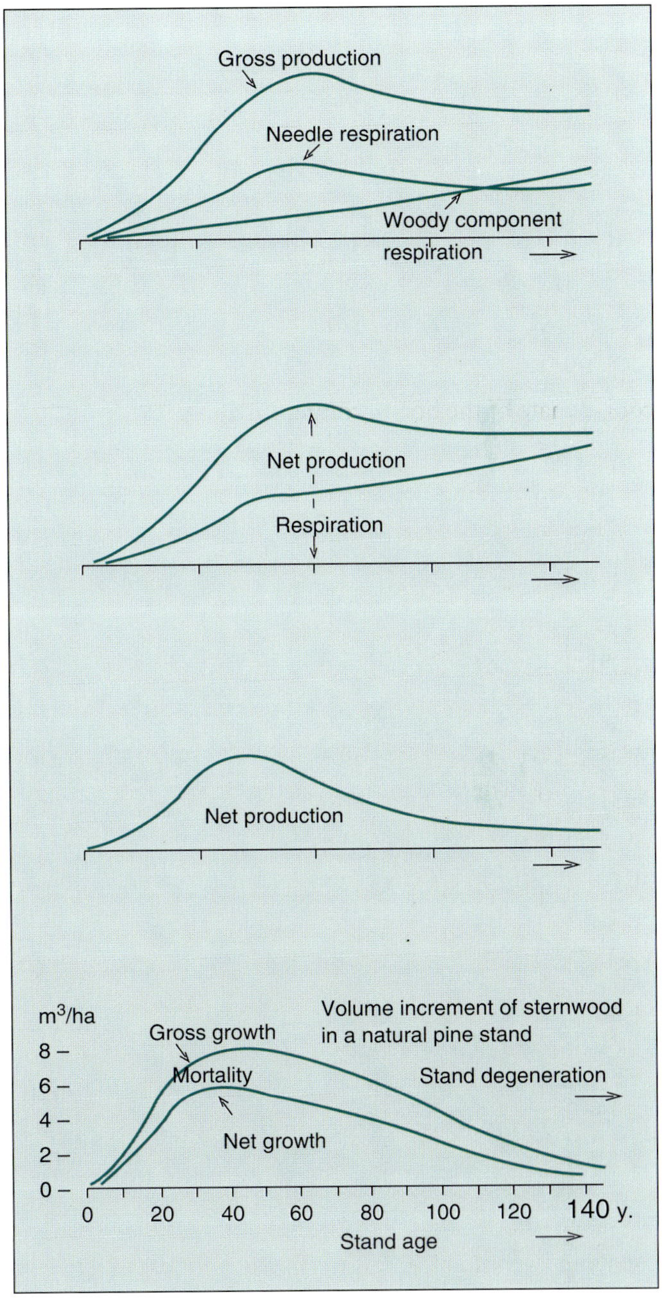

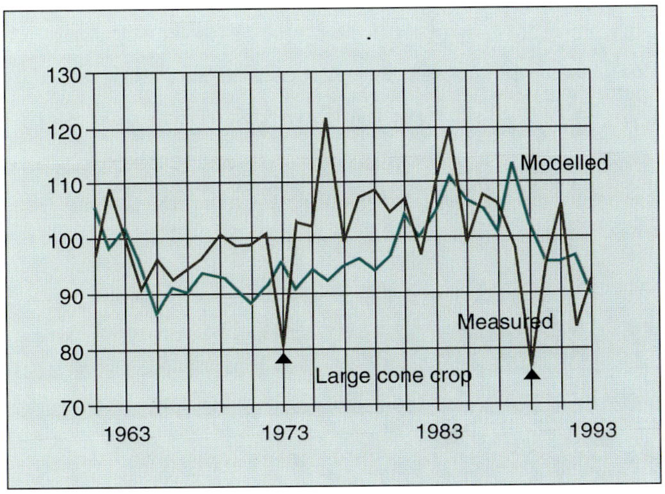

An index based on measured annual ring growth in spruce, and an index calculated from a model for temperature and precipitation during the growing season. The comparison shows that the higher the temperature and precipitation, the greater is tree growth. During years with large cone crops less of the biological production is utilised in stem growth.
Source: Kari Mielikäinen.

Fauna in the nutrient web of the forest ecosystem

Assimilating plants, being producers, form the first trophic level in the nutrient web of ecosystems. The second level comprises animals, chlorophyll-less plants and microbes, which decompose and use the biomass and remains of assimilating plants, as well as each other, as their food supply. The animals of the third trophic level consume other animals. Large predators such as bears, wolves and wolverines, and hawks, eagles and owls, form the top level of the food web.

Decomposing animals, which participate in the nutrient cycles and help to maintain high biological production, are an essential part of forest ecosystems. By decomposing plant and animal remains, they release nutrients for the assimilating plants. Most of them are rather small and live in the soil. However, maintaining their role as a part of forest biodiversity presupposes a better knowledge of the species involved and of the biotopes which favour them, and the development of forest management principles that take these factors into account.

Visible and identifiable animals have mainly been treated as damaging pests or game in traditional forestry science and in silviculture. Hunting and game management are a part of European forest management.

In Finland's climate insect damage is rather slight compared e.g. to the situation in North America. Insect damage mainly occurs locally and sporadically, and biological methods of pest control that replace chemical methods are increasing being used. Legislation concerning the storage of timber states that unbarked round timber should be stored in such a way that bark beetles do not spread into the surrounding forest, and that wood assortments are to be harvested from the forest during early summer before the bark beetles emerge.

Land and forest use, cattle farming and hunting have changed the natural relationships between animal species and resulted in an increase in species that are harmful for forestry. For instance, the almost complete disappearance of large predators and reduction in illegal hunting resulted in such a large increase in the moose population after the 1960's that they caused severe damage in young pine and birch plantations. The large reindeer populations in the reindeer husbandry areas of northern Finland have altered the ecosystems by reducing the coverage of lichens and retarding birch regeneration.

Afforested agricultural fields and ploughed forest regeneration areas are the reproduction sites of voles. When the vole population reaches peak levels there is a considerable increase in damage to young planted stands especially.

Successful silviculture and forest management presupposes a state of equilibrium between the tree-feeding animals and their natural enemies; the size of pest populations is kept low . Hunting and reducing the sites where pests can reproduce are other control measures. Regulated moose hunting has reduced the damage in seedling stands from its high level during the 1970's.

Nutrient cycling

Water and carbon dioxide, the starting materials of assimilation, are cycled between the atmosphere and the forest. Part of the nitrogen is also cycled via the atmosphere.

The main nutrients (macronutrients) are nitrogen (N), phosphorus (P), potassium (K), calcium (Ca), magnesium (Mg) and sulphur (S). The micronutrients are iron (Fe), manganese (Mn), copper (Cu), zinc (Zn), molybdenum (Mo), borate (B) and chloride (Cl). Plant tissue also contains silicon (Si), sodium (Na) which is essential for growth, and cobalt (Co) which is required by nitrogen-fixing bacteria.

Most of these nutrients are released from the weathering of minerals. Plants can only use nitrogen in the form of nitrate or ammonium. Nitrogen is deposited as nitrate and ammonium in rainfall and is released back into the atmosphere in forest fires especially. Certain plants such as grey alder (Alnus incana) and bog myrtle (Myrica gale) have bacteria in their root nodules which can fix nitrogen. Alder, which is a pioneer species, replenishes the nitrogen reserves in the soil.

When the growing conditions are favourable the soil contains nutrients in the optimum proportions required by plants. If any nutrient is in excess with respect to the others, the result is a nutrient disturbance. A shortage of plant-available nitrogen is the primary factor restricting plant growth in boreal coniferous forests. A lack of micronutrients causes growth disturbances, e.g. on drained peatlands.

On fertile sites the water transporting the nutrients flows freely through the soil. If the water is standing, tree growth is low even on fertile soils. Nutrients are transported by the water from the soil to the trees, and return to the ground in litter and in the remains of dead plants and animals. High site fertility presupposes efficient nutrient cycling which, in turn, requires rapid decomposition of the forest litter and the release of nutrients in a form available to plants. Calcium speeds up the decomposition processes.

The annual amount of forest litterfall, which consists of leaves, needles, flowers, cones, seeds and dead branches, is 1 500 - 2 500 kg/ha, depending on the tree species and site fertility. Whole trees also die and rot in dense, old stands.

In the initial stages of the decomposition processes the plant and animal remains are broken up into small pieces. Part of the nutrients are leached out at this stage. The soil animals, fungi and bacteria perform the final decomposition. Fungi are the most important group of decomposers in the boreal coniferous forests.

Decomposition and the release of nutrients are the faster, the higher the temperature of the soil surface, the greater the amount of oxygen, and the lower the soil acidity. Forest fire and the solar energy falling on treeless ground effectively speed up decomposition. Microbial activity presupposes sufficient nitrogen with respect to carbon in the litter. Carbon dioxide is evolved as a result of respiration by the decomposing agents. Nitrogen is bound up in the microbial biomass and in the humus.

Plant remains decompose much more slowly in a cold climate than in a warm one. Organic matter containing nutrients in an unavailable form does not decompose very rapidly and it accumulates in the humus layer in the boreal coniferous forests. The raw acidic humus in spruce stands especially slows down nutrient cycling and reduces the amount of nutrients available in the soil. An extreme example of this is the raw humus which develops in old spruce stands. The humus layer, together with the uppermost layers of the mineral soil, can contain as much as 4 000 kg nitrogen/ha, but less than 1% of this nitrogen is in a form available to trees.

Some nutrients are leached from the soil by percolation water and surface runoff. Leaching losses are low in boreal coniferous forests because the humus and mineral soil particles bind the nutrients and nutrient uptake is rather efficient. Erosion is not normally a problem. However, it does occur on sites with fine-textured soil lacking a humus cover, and where the slopes are steep.

The dynamics of the nutrient system are dependent on the development stage of the stand. Sites with a seedling stand contain plenty of nutrients and water, which are bound and cycled by the vigorously growing ground vegetation. As tree size and stand density increase, the ground vegetation gradually becomes suppressed and nutrients are freed for cycling between the trees and the soil. Binding by the ground vegetation reduces nutrient leaching after clear cutting.

As the stand ages, soil acidity increases, especially if spruce gradually becomes the dominant species, nutrients accumulate in the humus and site fertility decreases. In natural forests fire is important for renewing site fertility.

Nutrients which are mobile, e.g. nitrogen, phosphorus and potassium, primarily occur in the living parts of the tree biomass: leaves, needles, growing parts of the stem and branches and the fine roots. Nutrients such as calcium are stored in the dead part of the stem and branch wood.

The debarked stem of a tree, especially if there is a high proportion of heartwood, contains relatively few nutrients. Timber made from the stem contains so few nutrients that the nutrients removed in harvesting are replaced by weathering and nitrogen deposition from the atmosphere. Whole-tree harvesting, in which the crowns as well as the stem are removed, reduces soil fertility unless nutrients are replaced by fertilization. The production of the ecosystem can be maintained by leaving those parts of the biomass which are not suitable for timber to decompose in the forest.

THE "WARNING SPRUCE STAND"

A number of forestry colleges were established in different parts of Finland at the beginning of this century. In addition to training inspectors and other officials for the practical forestry sector, short-term courses were also arranged for forest owners. The Tuomarniemi Forestry College, near Ähtäri, was one of the first of these colleges.

The head of the college during 1909-1939 was Forest Officer Arvid Borg, who had extensive experience and strong views about forest management, forest fires and climax spruce stands. Ähtäri is located at an altitude of 150 - 210 m a.s.l. in the southern part of the Suomenselkä highlands in central Finland. The snow cover is thick and melts late in the spring. The high precipitation:evaporation ratio during the growing season promotes the formation of a thick moss layer in spruce stands and the paludification of forest land.

Borg established an experiment in a local forest in order to demonstrate the importance of forest fires in coniferous ecosystems, and called the experiment the "warning spruce stand". At the time the experiment was established in 1920, the warning spruce stand was an understorey spruce stand, growing on a site of the Myrtillus site type, that had been left after selection cuttings performed in 1908-09. A comparison site was clear cut in 1919-1920, burnt in 1920 and sown with pine and larch in 1922. The comparison stand was thinned in 1951. In 1987 when the comparison stand was 65 years old the stand parameters were as follows:

PARAMETER	WARNING SPRUCE STAND	COMPARISON STAND
Age, years	170	65
Stem number/ha	1 890	952
Dominant height, m	15	23
Mean diameter, cm	16	26
Volume, m^3/ha	151	313
Growth, m^3/ha/year	1,3	7,4

In 1915 Borg made the following entry in his travel diary: "How many thousands of forest owners in Finland have carried out precisely the same sort of measures in their own forests (i.e. left a poorly growing understorey of spruce), and with great satisfaction think that they now have a budding new forest! But this sort of "budding new forest" is really the end of the forest." - "The more correct treatment is to fell the spruce understorey at the same time as the overstorey, burn the logging residues and ground vegetation and sow pine seed in the ash if there are not enough seed trees left to restock a new stand. This results in a stand which starts to grow right from the ground surface, but in a couple of decades surpasses the original spruce understorey and is capable, above all, of reaching sawtimber size and not just producing poles." Borg's work in the district of Ähtäri has resulted in prescribed burning being used more than anywhere else in Finland.

In 83 years the spruce understorey has developed into a forest with a thick blueberry stand, dense moss cover, and Sphagnum peat growing in the depressions. The growth of the "warning spruce stand" has ceased and no new seedlings are being formed in the forest.

The 70-year-old pine stand in the comparison area was established by sowing on the burnt soil. The floristic composition of the ground vegetation is richer, tree growth is fivefold and the stem volume about double that of the "warning spruce stand".

Nitrogen content (% of dry mass) in different compartments of Scots pine. The stemwood harvested in cuttings accounts for 45% of the total biomass of the tree.

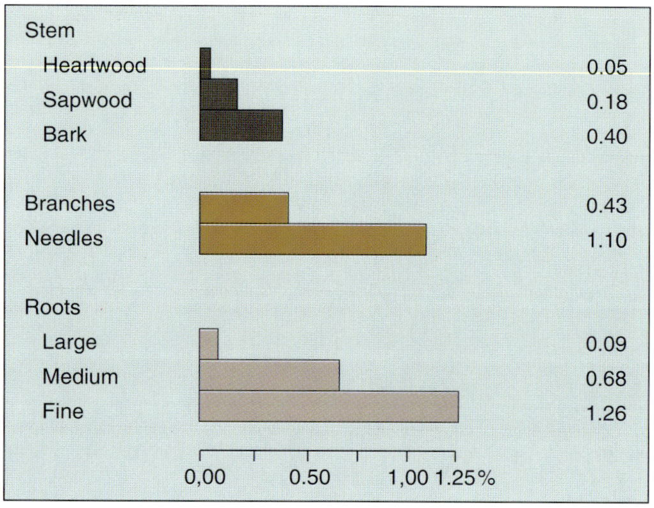

An artificially regenerated, closed spruce stand growing on a site of the Oxalis-Myrtillus *forest site type (OMT) results in an unfavourable microclimate in the stand. In the winter there is only a thin snow cover and frost penetrates deep down into the soil. In summer little light and heat penetrate down to the ground surface, retarding litter decomposition and decreasing nutrient availability.*

The presence of birch in spruce stands improves the microclimate. The soil in winter is protected by a snow cover. Mixed stands and the underlying soil are relatively warm during the growing season, and the easily-decomposable birch leaves promote nutrient cycling and prevent the development of a moss cover. The total yield of birch-spruce mixed stands is 5% greater than that of pure spruce stands.

Decomposition by the soil flora and fauna is faster in the temperate broad-leaved forest zone than that in the boreal coniferous forest zone. This prevents the development of a uniform moss cover and incompletely decomposed humus. The photograph was taken in a beech forest near Gothenburg in southern Sweden.

Most of the nutrients bound in the tree stand are located in the needles and bark, from where they are rapidly released into the nutrient cycle. The arctic hare (Lepus timidus) and bank vole (Clethrionomys glareolus) eat the bast at thin-barked points on the pine trunks during the winter. Bast was still used by people as an emergency food source during the 19th century.

The forest as a carbon store

In forest ecosystems, the tree stand, the biomass of the other plants and the soil represent a major carbon store. Fresh stemwood contains more than 50% water. The dry matter contains about 51% carbon. The amounts of dry matter vary by tree species. The mean values are:

Pine	405 kg/m^3
Spruce	385 kg/m^3
Birch	476 kg/m^3

In addition to the stemwood, the branches, needles, leaves, roots and ground vegetation also contain carbon. Forest soil, plant litter and humus, and especially the peat of wetlands, contain relatively more carbon than the tree stand. It has been estimated that the topmost 60 cm-thick layer of soil on the 23.2 million ha of forest and scrub land in Finland, which includes 6.2 million ha of peatland, contains 2 798 million tonnes of carbon. When the trees and ground vegetation are included the total is 3 607 million tonnes.

The peat deposits in many wetlands are thicker than 60 cm. The amount of carbon which has been bound in the peat of wetlands since the last ice age is estimated to be 6 250 million tonnes.

Intensive forest management has increased the amount of carbon bound in the stemwood by 199 million tonnes, which is an increase of 33% since 1937. When the other biomass components are included the total is 398 million tonnes of carbon, which corresponds to an accumulation rate of 7.5 tonnes per year.

The capacity of the forest and the tree stands to bind carbon, coupled with the cultivation and use of wood as a part of the natural carbon cycle, can be used to regulate emissions of carbon dioxide into the atmosphere and to maintain the carbon dioxide balance of the atmosphere as a part of sustainable development.

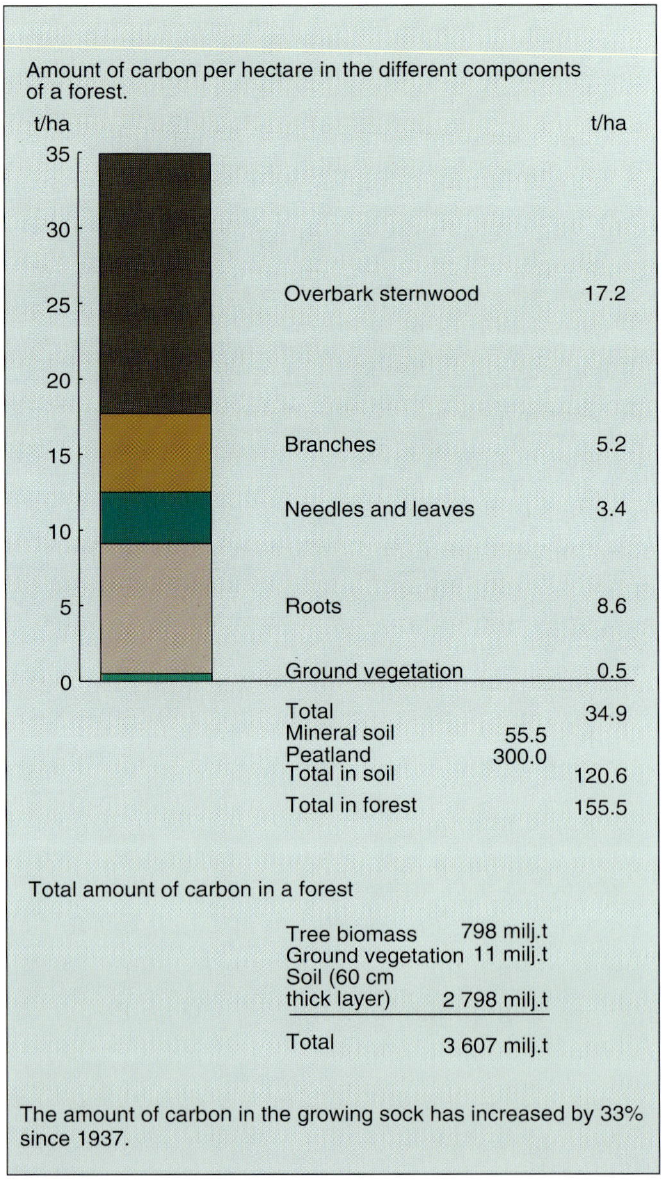

■ *Amount of carbon stored in a forest* (1990). *Source:* Karjalainen, 1991.

NATURAL FOREST

Afforestation of the land after the last ice age

The properties of the tree species in the boreal coniferous forest zone have developed under conditions characterised by great climatic variation and recurrent ice ages. The cooling climate and expanding continental ice sheet have forced the trees, other plants and animals to move to the south and south-east. They have returned from there during the warm periods between the ice ages.
The number of species was reduced during this backwards and forwards migration because the seas, large lakes and mountain ranges run diagonally across the migration path. Pine and spruce are the only coniferous tree species which naturally form forest in Finland and the other Nordic countries.
The trees of the boreal coniferous forest have become adapted to climatic variation and to form forest on treeless ground. The plant communities which are dominant in this zone are amongst the most hardy in the world.
The last time the continental ice sheet spread to the south it pushed out all the plants and animals and carried with it the organic soil and much of the overburden. Some of the present fell species may have spent the last ice age in refuges on the Atlantic coast in the western and northern parts of Norway.
The first areas to be exposed by the melting ice sheet were located between the Kola Peninsula and Lake Ladoga. Relict plants from this migration period can be found in the river valleys in Salla and Kuusamo, along the Maanselkä highland and in parts of Karelia near Lake Ladoga.
The ground which was released from its cover of ice and water consisted of a range of soil types. There was no longer any humus and most of the soil was unweathered. A living soil suitable for more demanding plants was gradually formed from the remains of lichens, algae, mosses, club mosses, other plants and micro-organisms, and mineral soil particles. Dwarf shrubs, sedges and grasses joined the first group of pioneer species in the patchy vegetation. If there were seed-bearing birches in the vicinity, birch was the first pioneer tree species. 12 000 - 11 000 years ago most of the peninsula of Finland was covered by ice and meltwater. Birch was the major tree species on land rising out of the sea in the eastern parts of Finland and in Karelia along the coast of the White Sea. Isolated specimens and groups of pine, and perhaps some grey alder and aspen, were to be found among the birches. Grey alder occurred along the watercourses flowing off the melting ice sheet.
The pre-boreal period started 10 000 years ago when about one third of the Finnish Peninsula in the south east had been freed from the ice and water. Birch was the dominant tree. Elm (Ulmus), black alder (Alnus glutinosa) and hazel (Corylus avellana) were an admixture on suitable sites with the pine, grey alder and aspen that had arrived earlier. The climate was relatively warm judging by the remains of water chestnut (Trapa natans) found in lake sediments in northern Savo. Nowadays the northernmost area where water chestnut occurs is Latvia. Most of the species found in today's forests became established at the end of this period.
The boreal period started 9 000 years ago. The climate was initially dry, warm and slightly continental. Pine became the dominant tree. The forest fires promoted by the climate encouraged the spread of pine. New plants arrived from the south-east, as well as from the south over the Gulf of Finland. Birch spread to the west and north and reached its widest distribution about 8 500 years ago.

The climatic and historical eras since the last ice age and changes in the tree species composition. Sources: Eronen, 1991, Kalliola, 1973.

Post-glacial climatic and historical periods

	Before the present	Period	The Baltic	Climate	History	Pollen composition
Post-glacial perod	1 000					
		Sub-Atlantic	Limnean Sea	Cool, humid	Iron Age	Few
						hardwoods
	2 000					
						Norway spruce
	3 000		Litorinan Sea			becomes comm.
						Scots pine
				Cooling down,	Bronze	becomes comm.
	4 000					
		Sub-Boreal		humid	Age	Norway spruce
						appears
	5 000		Ancylus Lake			Abundant
						birches and
						hardwoods
	6 000					
		Atlantic		Warm, humid		
					Stone Age	Alder becomes
	7 000					common
				Warm, dry		
						Abundant
	8 000					
		Boreal		Rapidly warming		Scots pine
	9 000					
				Arctic, cool dry		
		Pré-Boreal				Abundant birch
	10 000		Yoldian Sea	Sub-Arctic,		
		Younger Dryas		warm, dry		
Late ice age	11 000					
		Alleröd				Herbs and
				Arctic, cold,		dwarf shrubs
	12 000	Older Dryas	Baltic ice lake	dry		
		Bölling				
	13 000					
	14 000					

The conditions were favourable for species growing on the sandy eskers, rocky slopes and groves.

During the Atlantic period 8 000 - 5 000 years ago the climate was warm, moist and maritime. The forest in southern Finland became dominated by broad-leaved species, primarily birch. There was also a lot of pine and grey alder. The distribution of black alder, elm, ash (Fraxinus excelsior), oak (Quercus), lime (Tilia), maple (Acer) and hazel was greater than at any other time during the post-glacial period. Elm was the hardwood species which spread furthest to the north; its remains have been found in Kainuu. The first spruce reached the easternmost parts of Finland.

Southern and maritime species arrived from Estonia and Sweden, e.g. yew (Taxus baccata). Most of the present-day south-western archipelago was covered by water, and the sea was a barrier to the spread of southern and south-western species. The vegetation zones of today were located further to the north. Judging by the trees excavated from the bottom of mountain lakes, the timber line on the fells in Lapland was about 200 m higher than it is today.

During the sub-boreal period 5 000 - 3 500 years ago pine spread throughout the whole country and spruce became common in the east. Hardwood species, apart from oak and black alder, became less common. Elm, which regenerates from sprouts, was better able to maintain its position compared to the other hardwoods. Pine-dominated stands spread to the shores of the Arctic Ocean. New ground vegetation species typical of the taiga arrived from the south-east and east.

Spruce became widespread in eastern Finland about 5 500 years ago and spread to the west, north-west and north. There were appreciable amounts of spruce in south-west Finland and Lapland about 3 000 years ago. Spruce attained its greatest distribution in northern Finland 2 500 - 2 000 years ago, and reached the Åland Islands after the start of the Christian era. Spruce is still spreading naturally in the western parts of Scandinavia.

It took about 1 000 years after spruce first appeared for it to become common over most of the country. There were presumably a few spruce in eastern Finland and in Petsamo already during the late early stages of the post-glacial period. The present-day distribution of spruce is often attributed to variation in the climate. Another important factor is the frequency of forest fires and spruce's dependence on specific types of soil, as well as the climate during each period.

Spruce is easily damaged and killed by forest fires. When the period between fires is short, spruce does not have time to occupy those sites where it is biologically stronger than other tree species. Spruce maintains its dominant position along streams in the vicinity of the timber line and on some types of swamp.

The sub-Atlantic period started about 3 500 years ago. During this period the climate was cool, moist and relatively maritime. The present-day plants of mineral soils and peatlands consolidated their position. Spruce became the dominant tree species in climax stands, apart from those on the most infertile sorted sand and gravel soils, on exposed bedrock and on infertile peatlands. The areas where demanding southerly species grew shrank and became fragmented. Man's influence on forest ecosystems and the relationships between species became stronger than that of nature.

REMNANTS OF THE ICE AGE IN JOSTEDAL IN CENTRAL NORWAY.

A cool climatic period at the beginning of the 1550's resulted in the expansion of the glaciers at Jostedal in the mountains of central Norway. The glacier started to flow off Jostedalsbre and destroyed settlements in the valleys during 1680-1750. The Little Ice Age reached its peak around 1750, after which the glaciers retreated back up the valleys to their present position. Jostedalsbre is a so-called ice-cap glacier and is located approximately halfway between Bergen and Trondheim. It is the largest glacier in Continental Europe. It covers 48 600 ha and its peak reaches an altitude of 2 000 m a.s.l.
More than ten glaciers run down into the valleys off the central glacier at Jostedal, and Nigardsbre is, owing to its accessibility, the most well known. The glacier extends down the U-shaped valleys at a level clearly above that of the timber line. At its greatest extent the glacier reached about 5 km lower down the valley of Nigardsbre. During the Little Ice Age the mean temperature was about 1°C lower than it is today. Between 1886 and 1940 the temperature in the Bergen area has increased 0.4°C. The glaciers that have retreated back up the valleys have left bare ground to be colonised by plants and animals.
Birches, grey alder and pine arrived in Jostedal already 8 000 - 9 000 years ago. Pine formed a real forest cover about 8 500 years ago. The land exposed by the retreating glaciers is nowadays dominated by broad-leaved forest, and birch and alder are clearly more common. Birch has immediately invaded the exposed ground and alder has gained a foothold in the earlier flat sections of the glaciers where layers of fine-textured soil have accumulated. Pubescent birch is clearly more common than silver birch. Small groups of aspen are rather common on the more fertile sites. There are small stands of hazel (Corylus avellana) and elm (Ulmus) on the warmer southern slopes. Norway spruce has not yet spread naturally into the area, and the present spruce stands have been planted.
The Nigardsbre valley is located at an altitude of 260 - 300 m a.s.l., and the edge of the glacier is about 330 m higher up. The way in which different species of plant and animal colonise land released from the glacier has been studied in the valley. There is a rich flora in the area comprising 430 different vascular plants, most of which are relatively rare in the area.

Polished bedrock and gravel become exposed as the glacier retreats. The small river flowing out from under the glacier carries mineral nutrients into the glacial lake and delta that are often formed.

Birches rapidly colonise the ground released from the ice because they produce seed crops every year. Birch is often the first plant to appear on bare gravel soil.

The glacial river transports large amounts of mineral nutrients down into the valley and the plants and animals colonise the new habitats formed along the river bank as the glacier retreats. The edge zones of the glacial streams form ecocorridors for the plant and animal species.

Grey alder (Alnus incana) is the most competitive tree species in the water-logged depressions containing fine-textured, fertile soil. Alder reaches maturity at an age of 40 - 50 years, after which the stands starts to degenerate naturally. If spruce has managed to spread into the area it occupies the growing space freed by the dying alders.

The ability of Scots pine (Pinus sylvestris) to colonise exposed mineral soil is clearly inferior to that of birch. However, individual trees are able to gain a footing. Being a long-lived tree, pine gradually occupies the sites previously covered with birch.

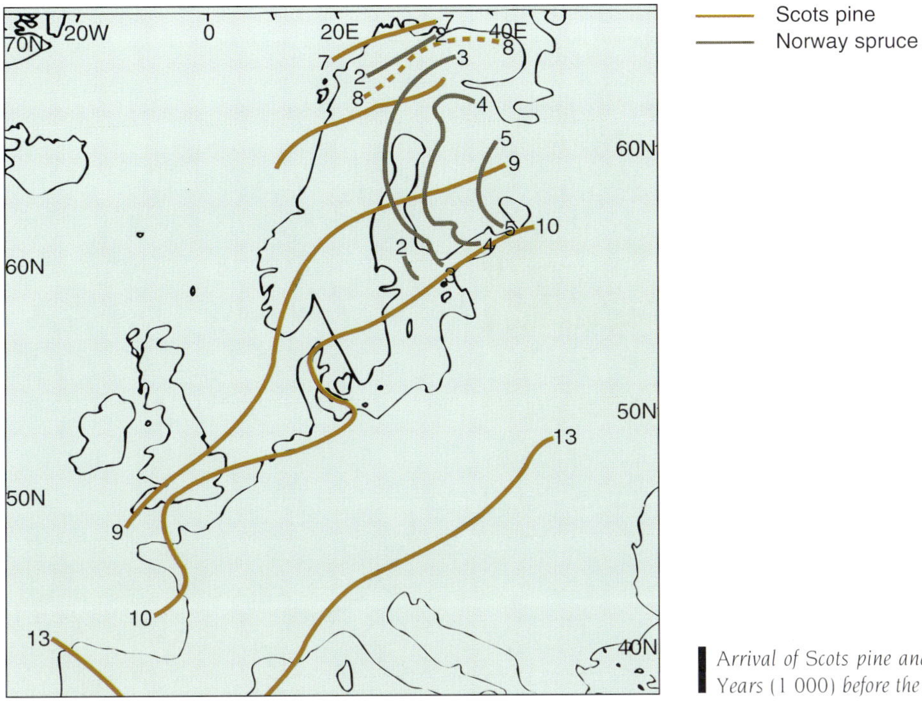

Arrival of Scots pine and Norway spruce in Finland after the last ice age. Years (1 000) before the present. Source: Korhola, 1990.

Bird's foot trefoil (Lotus corniculato) is a member of the legume family. It grows together with lichens on infertile gravely soils. Both plants fix nitrogen and thus improve the soil for other plants.

The valley contains curved terminal moraines marking the previous positions of the glacier's edge. The Salpausselkä eskers in southern Finland were formed in the same way at the end of the last ice age. However, they are much larger because the continental ice sheet remained stationary for extended periods.

AFFORESTATION OF LAND EXPOSED BY LAND UPLIFT

Following the last ice age the land has been rising up along the coast of the Gulf of Bothnia at a faster rate than anywhere else in the Baltic. During the past 3 500 years it has averaged one meter a century, but the overall trend is gradually declining. Nowadays it is about 90 cm a century in the narrowest parts of the Gulf of Ostrobothnia. Uplift is slowest in the south-eastern parts of Finland - about 20 cm a century.

The properties of the soil released from the sea and changes in the soil as it "ages" and becomes covered with forest have been studied rather intensively. Right at the edge of the shore the pH of the soil is almost 7 owing to the presence of sea water, but further inland where the soil has been exposed for a longer period soil acidity increases. Close to the sparse willow zone and outermost grey alders it is already about 5.5.

Grey alder (Alnus incana) is the first tree species to form closed forest on land rising out of the sea. This species can withstand temporary inundation by sea water during seasonal high tides. The first alders appear on the shore zone within 50 years after the land has become exposed. However, the arrival of the species is greatly dependent on variations in high water mark, the type of soil and the wind conditions. The site becomes suitable for conifers within 120 - 150 years, and the first spruces appear when the soil has risen 120 cm above the average sea level. Grey alder adds nitrogen to fine-textured soil, making it suitable for demanding species. Another nitrogen-fixing species is sea buckthorn (Hippophae rhamnoides) which grows in the archipelagos.

Erik Appelroth, a Finnish forester and scientist, started trials with different tree species in the alder zone on land rising up from the sea near Uusikaupunki. He also investigated changes in the fertility of uplifted soil. After the alder stage, the site quality gradually changed from fertile to rather infertile forest site types: grovelike (Oxalis-Myrtillus), fresh (Myrtillus), dryish (Vaccinium) and finally dry (Calluna) and even lichen (Cladonia) site types (cf. pages 144-147).

Appelroth cleared primary alder forest and replaced it by planting with spruce. A sample plot was later marked out in the stand. The comparison site was a naturally regenerated grey alder stand with a stem volume ranging from 20 to 70 m3/ha. According to the results, the increment of the spruce stand growing on land fertilised by the alder stand has been faster, during the first 50 years, than that of spruce growing on the most fertile site type (Oxalis-Myrtillus). The total yield was 415 m3/ha and mean increment 8.3 m3/ha/year. Although the early development of the stand has been fast, the height growth has slowed down with increasing age. The nutrient reserves of the fertile soil have decreased with increasing stand age, and continuous land uplift has had a detrimental effect on the soil moisture conditions. The results of Appelroth's experiment are clearly visible in the coastal forests in the land uplift zone. Soil fertility decreases as the land rises. Commercial spruce plantations have been established along the Ostrobothnia coast on land cleared of grey alder.

Appelroth's primary stand is the comparison area with grey alder (Alnus incana). The tree stand is of poor quality and degenerating, and its stem volume is about 60 m3/ha. Grey alder reaches a maximum age of about 45 years along the coast of the Gulf of Bothnia, and it has a strong self-thinning ability.

The 59-year-old spruce stand, planted by Appelroth, *in* 1993. *According to the vegetation cover, the area is a herb rich mineral soil site (*Oxalis-Myrtillus *forest site type). The growth of the spruces has declined compared to the average growth level for spruce on this type of site.*

This planted spruce stand has grown well under the primary grey alder stand in the shore zone. The alder overstorey has been felled to provide space for the spruces to grow.

Wave action sorts the sandy soil in the shore zone into different fractions, the finest material collecting in depressions. After land uplift these depressions subsequently support more fertile vegetation, while elevated areas correspondingly have less fertile vegetation. The birches have already degenerated. Spruce gradually invades the site as the ground rises and its site quality index has become too poor for spruce.

The soil on land rising up out of the sea is young and has not yet been modified by the movement of water down the soil profile. There are no detectable horizons because podsolisation has not had time to have an effect. It takes about 1 000 - 1 500 years for a podsol soil horizon to develop on the Ostrobothnian coast. Precipitated iron has given a reddish-brown tinge to the soil. Chemical analyses have shown that a podsol profile can develop within 200 - 300 years even.

SUPRA-AQUATIC LAND AFTER THE LAST ICE AGE

During the period following the last ice age most of Finland was covered with water. However, the highest hills in central Finland, the hilly areas of eastern Finland, the Kainuu highlands and the northernmost parts of Lapland remained above the water level (supra-aquatic land).

The Baltic glacial lake covered the southern coast of Finland 10 500 - 10 200 years ago, the rest of the country still being covered by ice. The Yoldian Sea phase occurred about 10 000 years ago when the glacial lake broke out into the Atlantic through central Sweden. Land uplift continued and, when the sea connection was severed, the Ancylus Lake was formed about 9 000 years ago. At that time most of southern and western Finland were under water. The present connection to the North Sea opened about 7 500 years ago through the Danish Straits.

The influence of the lake or sea is reflected as paludification, in soil fertility and as washed-out shoreline boulder fields, which are called "devil's fields" in the countryside. The land above the shoreline consists of unsorted glacial till soils, the banks are stony and the lower parts of the slopes comprise sorted layers of sand. The earliest settlers in eastern and northern Finland found the most fertile sites for slash-and-burn agriculture and fields on the unsorted hilltops. They were also favourable because summer frost did not destroy the new shoots of their crops.

The highest ancient shores in Finland are located at an elevation of more than 200 m a.s.l. near Rovaniemi in Lapland. The photograph shows a stony field washed out by the sea, which the locals call "shingle", located in the Pisavaara Nature Park. It was formed about 9 000 years ago.

There are a number of hills in northern Satakunta whose tops lay above the Yoldian Sea and were not subjected to the sorting effect of water. The soil contains 12 % fine material (silt and clay) and is therefore rather fertile. The spruce stand on Alkkianvuori Hill is growing on a site of the Myrtillus forest site type and is located 190 m a.s.l.

The lower part of the ancient shore on Alkkianvuori Hill has been washed out by water and the site is of the Calluna forest site type. The soil contains only 2 % fine material and is rather stony. Wave action has resulted in a fertility decrease equivalent to two forest site type classes compared to the spruce stand.

Paludification

A wetland is a community of wetland plants that form peat. Peat is formed in the anaerobic conditions below the groundwater table of wetlands as a result of incomplete decomposition of the plant remains. Close to the surface where the water contains oxygen the plant mass decreases by 80 - 95%. In anaerobic conditions the corresponding reduction is 5 - 10%. Anaerobic bacteria decompose part of the peat into methane and carbon dioxide. Humified peat is primarily composed of organic material, the mineral content being at the most 5% of the dry weight.

A mire forming peat is the most common form of wetland in the cool, moist conditions of the boreal coniferous forest zone where evaporation is less than precipitation for most of the growing season. Other forms of wetland include land periodically covered with floodwater and the land rising up from the sea along the coasts.

Paludification has been regulated to a great extent by the varying rate of land uplift in different parts of the country after the ice sheet retreated, and changes in the macroclimate. During cool, moist periods paludification was faster than that during warm, dry periods.

The onset of paludification on mineral soils presupposes standing or slowly moving water that results in an increase in water-absorbing Sphagnum mosses and other wetland plants. Part of the flat land lifting up from the sea became covered with wetland vegetation immediately after it passed out of the reach of the waves and ice. The earth's crust was originally depressed by the weight of the continental ice sheet, and is still rising at a rate of 9 mm/year on the Ostrobothnian coast in the narrowest part of the Gulf of Bothnia. Land uplift decreases on moving from the north-west to the south-east, and is 2 mm/year on the south-east coast of Finland. The land rising up from the sea in the Gulf of Bothnia is flat and the rivers running through the area often flood extensively in the spring.

Forest fires have caused paludification on mineral soils. After destroying the trees the ground water rises up close to the ground surface owing to the lack of evapo-transpiration, and wetland plants become common. The dams built by beavers also raise the water level on mineral soils and cause paludification.

A relatively high proportion of the peatlands in southern Finland have been caused by the gradual overgrowth of lakes, pools and shallow sea bays. This occurs in two ways. Paludification takes place via the bottom in relatively nutrient-rich, shallow watercourses, where the water level falls or the bottom rises owing to the gradual accumulation of fine-textured material. Wetland vegetation spreads along the bottom and gradually displaces the water.

When paludification takes place via the surface of the water, Sphagnum mosses form a raft, extending from the shore over the surface of the water, and the water is gradually displaced. This occurs in nutrient-poor, relatively deep and steep-shored waters and in sheltered inlets.

A peatland stops spreading when it has reached the greatest expanse permitted by the terrain. Its depth increases at a rate of about 2 mm/year. The increase in thickness of the humidified peat layer is smaller than this.

The plants in some peatlands obtain their water and nutrients only from rainfall. Such sites are the Sphagnum moss, raised (ombrogenic) bogs where the trees are short and bushy or completely absent. Below the nutrient-poor Sphagnum peat there are layers containing more nutrients. The earlier development stages have included peatland types with a tree cover. As the peat layer grows the wetland species often spread to the surrounding mineral soil.

Trees grow best on fertile peatlands where there is flowing water and receive their nutrients from water flowing in from the surrounding mineral soils. Trees also grow moderately well on thin-peated soil where the tree roots can penetrate down to the underlying mineral soil.

In forestry the biological definition of a peatland is a site where the coverage of wetland species, primarily Sphagnum mosses, is at least 75%. According to the geological definition, a peatland has a peat layer that is at least 30 cm thick. In earlier days peatlands in the biological sense covered one third of the land area of Finland, i.e. about 10 million ha. Of this about 1 million ha have since been converted into agricultural land. There are 9.0 million ha of peatland classified as forestry land, which is 34% of the total forestry land area. The proportion of peatlands is highest, 60 - 70%, in the low-lying northern parts of the Gulf of Bothnia, and over 30% in most parts of northern Finland and below 10% in the south.

Ombrotrophic raised bogs are to be found in the southernmost parts of the country below a line stretching from Vaasa to Joensuu. The centre of the bog usually consists of a mixture of wet hollows and dry mattocks, and is normally higher than the surrounding bog. The treeless central part receives mineral nutrients from rainfall only, and it is poor in nutrients like Fuscum bogs or Fuscum pine swamps. The peatland is surrounded by forested peat bogs and spruce swamps. The photograph shows Pyrinsuo bog, near Tammela in southern Finland.

Aapa bogs occur from the northern edge of the ombrogenic bog zone right up to the northern timber line. Aapas are extensive, treeless bogs which obtain their nutrients from the spring floods. There are hummocky ridges running at right angles to the direction of water flow in the bog. The wetland site types are open oligotrophic bogs and eutrophic fens. Palsas, or peat hummocks, with a permanently frozen central part, occur in the northernmost aapa bogs. Palsas are formed along the northern edge of the pine forest. The photograph shows an aapa bog with numerous palsas. The bog is located on the spruce timber line in the Pallastunturi National Park.

The bare soil exposed by land uplift starts to paludify in protected inlets and bays where the variation in the sea level is not great. Sphagnum mosses start to grow in the reed or sedge zones along the Ostrobothnian coast and gradually form a layer of peat. Primary peatlands account for about 40% of the area of over 8 000-year-old inland peatlands, and about 60% of the area of younger coastal peatlands.

The ground water table may rise, e.g. after a forest fire, and the hair mosses (Polytrichum spp.) and Sphagnum mosses which appear at a later stage start to form a swamp. Most cases of forest land paludification originate from a forest fire. The photograph was taken in southern Finland.

Sphagnum mosses grow along the shores of small lakes and ponds; this is the starting point for the infilling and paludification of the watercourses. The mosses expand simultaneously over the surface and bottom of the lake or pond. The peat mat gradually spreads towards the centre of the pond as the sludge accumulates on the bottom. The most important factor in the formation of raised bogs in southern Finland is the infilling of watercourses.

Tree species in a changing climate

The last continental ice sheet covered the whole of Fennoscandia and extended to the eastern and south-eastern side of the White Sea, Lake Onega, Lake Ladoga and the Gulf of Finland, to the southern side of the Baltic Sea and North Sea, and to the British Isles. The ground was permanently frozen right down to the Alps. The climate was coldest 20 000 - 18 000 years ago. On the basis of plant remains it has been estimated that the summer mean temperature was 4 - 5°C lower in the maritime region of France and 10 - 5°C in the central parts of the continent than it is today. The tree species of the boreal coniferous forest zone had retreated to the mountain ranges of southern Europe.

The world's climate warmed up about 15 000 years ago. The continental ice sheet retreated to the north and north-west, with the trees following close behind. The climate cooled down rather abruptly about 13 000 years ago. The annual mean temperature fell by 5 - 6°C within a few decades. The retreat of the continental ice sheet was interrupted in south Finland. The tundra recaptured the land from the trees which had already reached Central Europe.

The Ice Age is estimated to have ended 11 000 years ago. The annual mean temperature started to increase 9 000 years ago at a time when the whole of Finland had become exposed. The annual mean temperature was about 2°C and the summer mean temperature 1 - 2°C higher than 6 000 years ago. The climate cooled down to its present level around 2 500 years ago.

Cold periods are usually drier than average. The climate was relatively humid during the warm Atlantic period. Precipitation increased during the cool sub-Atlantic period, when a reduction in evaporation increased paludification.

More recent significant changes were the warm period at the end of the Middle Ages. After that the climate cooled down into a "minor ice age", which started halfway through the 16th century and ended at the beginning of the 19th century. At its coldest the annual mean temperature was 1 - 2°C lower than at present. The annual and seasonal variations in temperature were great. The timber line receded in the north and field cultivation had to be abandoned over extensive areas.

The climate was exceptionally cold during the 1860's. The harvest failed. Famine drastically reduced the population. After this the climate again warmed up. The mean temperatures in the warmest years during the 1930's were, calculated as five-year sliding averages, 2 - 3°C warmer in Helsinki than the coldest periods during the last century. The annual mean temperature has remained the same or decreased slightly since the 1930's, and the annual variation has been large. Thus the warmest mean temperature during the 1930's was 5.5°C higher in the coldest years during the 1960's. The turning points in winter and summer occur earlier. Precipitation appears to have increased in northern Finland and slightly decreased in southern Finland.

Our knowledge of climatic changes and the structure of the vegetation shows that the trees of the boreal coniferous forest zone have become adapted to withstand relatively large changes in temperature. As the climate cools down the competitiveness of conifers compared to broad-leaved trees improves and the timber line recedes. The effect of climate warming is the opposite. Overall there have been little changes in the tree species composition.

Tree species and their properties

Tree species can be divided into those forming forests and other species. The former are the natural species of the boreal coniferous forest zone. Their proportion out of the volume of the present standing crop is 99.5%.

The other species are primarily broad-leaved trees of the more southerly mixed forest zone. They grow solitarily or in groups in the south-western archipelago and along the coast. In our present climate they have to be protected against competition from the dominant tree species and regenerated by planting. Exotic tree species are cultivated in field experiments.

The forest-forming tree species

In tree species dynamics the tree species are divided into pioneer and climax trees. All the broad-leaved trees are pioneer species, and spruce is the only clearly climax tree. Pine also forms forest on treeless land, and it can be considered to be a climax species on some types of site.

There are two species of birch, silver birch (Betula pendula) and pubescent birch (B. pubescens), which can intercross.

Silver birch grows best on clay and silt glacial tills, as well as on spruce swamps where the peat layer is thin and the water flowing. It attains a larger size than pubescent birch, and is a valuable raw material for plywood. Pubescent birch is biologically stronger on moist upland sites and on peatlands, as well as in northern Finland.

Finland's tallest endogenous tree species.

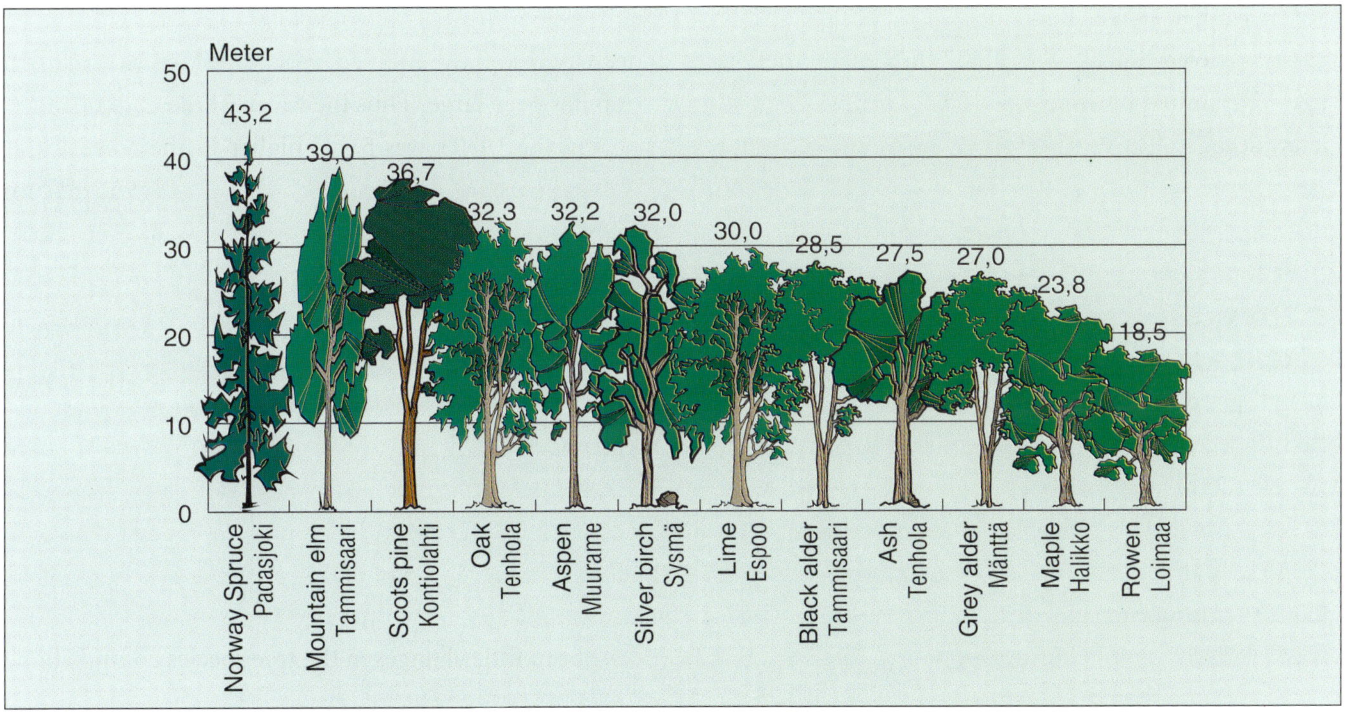

Birch produces, already at an early age, large amounts of light seed which are spread over wide areas by the wind. The seeds of the silver birch germinate readily and the seedlings grow best on exposed, burnt, moist mineral soil. Burn-and-slash agriculture increased the proportion of this species. Pubescent birch regenerates the best on moist mineral soils and on peatlands. Birches reproduce from sprouts and rapidly form forest on sites where the conifers have been removed by felling or destroyed by fire.

The seedlings require light and plenty of growing space. They are not sensitive to frost. Drought, however, kills the young seedlings. The seedlings and saplings grow rapidly, those developing from sprouts initially at a faster rate than those growing from seed. The growth of sprouts slows down at an early stage and they are usually affected by rot. Young birches grow faster than conifers of the same age, and kill off the pines growing under them.

The growth in height and stemwood volume of birch stands is greatest at the age of 20 -40 years in southern Finland, and somewhat later in northern Finland. Growth slows down at an earlier age than in conifers. Height growth ceases and birch stands start to degenerate after about 80 - 90 years. The height of the largest birches is less than that of conifers growing on the same type of site. Volume production is lower than that of conifers, but the density of the wood is greater. However, when the density is taken into account, the production of dry mass by birch is approximately the same as that of conifers.

The self-thinning capacity of a birch stand is good, and in the case of silver birch better than that of pine. The trees require a lot of growing space in order to maintain a full crown and produce a thick trunk. Birch does not recover very well from a dominated position.

The root system of birch is strong and extensive; it penetrates deep down into the soil and holds the tree firmly in the ground. It is rather resistant to windthrow. The bark on the butt of the stem of old birches, especially silver birch, is thick and intensively suberised. Old birches are relatively resistant to forest fires and retain, even when damaged, their ability to produce seed.

The snow bends and breaks the trunks of birches in overstocked young stands. Snow and other agents which damage the trunk and branches increase their susceptibility to infection by rot fungi. Moose, voles and hares can cause extensive damage in young birch plantations where the number of stems is low compared to that in natural stands. Overlarge reindeer populations in the north of Finland retard birch regeneration.

Birch is susceptible to rot fungi, which spread from damage points on the stem, branches and roots. Root rot reduces the capacity of old birches to withstand strong winds.

Aspen (Populus tremula) is common, but its share out of the total stem volume of the standing crop is very low. The most favourable sites for aspen are to be found in the southern parts of the taiga zone in European Russia. It cannot compete very well with other tree species under the rather severe climatic conditions in Finland. The formation of tall, thick stems requires fertile calcium-rich, moist soils with a high clay and silt content.

Aspen produces a tremendous number of light hairy seeds which the wind carries over long distances. High seed production starts at an age of 30 - 40 years. However, aspens that have grown from seed are rather rare owing to the fact that germination of the seed and development of the germling requires mineral soil free of ground vegetation and humus. The presence of aspen in our forests is

Swiss mountain pine (Pinus mugo) is one of the tree species, adapted to the Alps and central European mountains, which has not spread to the north. However, the species grows well in Scandinavia right up to Lapland. The photograph shows a subspecies of Swiss mountain pine (Pinus mugo subsp. uncinata) in the Engadin National Park in eastern Switzerland. It has a straight trunk and regenerates in small openings made by wind and insect damage; the seedlings have a bushy form.

The Sätkänävaara silver birch (Betula pendula) stand is located near Kittilä almost 150 km to the north of the Arctic Circle. It is one of the northernmost pure silver birch stands, and regenerated naturally about 110 years ago after a forest fire. The stem volume of the stand, managed as a field trial by the Finnish Forest Research Institute, is 70 m3/ha, increment about 1 m3/ha/year, dominant height 17 m and mean diameter 23 cm. The stem number is about 250 trees/ha.

primarily due to its ability to produce large numbers of fast-growing root suckers. The germlings are sensitive to frost, drought and insect and fungal damage. The leaves, shoots and bark of aspen are preferred by moose, voles and hares.

The seedlings grow extremely fast. The growth of sprout seedlings slows down earlier than in birch. Aspen stands reach biological maturity at an age of 50 - 60 years. Aspen maintains habitats for its own characteristic flora and fauna, and increases biodiversity.

The root stock is strong. The sprouts formed by superficial roots spread over a wide area. The bark is thin on young trees, but relatively thick on the butt of old specimens. Aspen is susceptible to rot fungi, which shorten its life-span and reduce its resistance against wind.

Grey alder (Alnus incana) is one of the first trees to start forming forest on treeless ground. Slash-and-burn agriculture and forest grazing have increased its proportion. It is a strong pioneer species on abandoned agricultural land and on land rising up from the sea. The root nodules of grey alder contain bacteria which fix annually atmospheric nitrogen at levels of 10 - 60 kg/ha, depending on the stand density and site fertility. Grey alder produces tall, straight stems only on the most fertile sites. On sites of low fertility and in the north it is a bushy tree. It reproduces through seeds and forms a large number of sprouts. The seed crop is large, but most of the seeds die as a result of drought and competition from the ground vegetation. The stands of alder produced from sprouts on fertile sites are usually so dense that spruce seedlings cannot grow under them. The seedlings withstand direct sunlight and frost. Grey alder grows fast when young and starts to produce seed at an age of 5 - 15 years. Growth slows down earlier than in other tree species. Height growth ceases and alder stands start to degenerate at an age of 40 - 50 years.

Grey alder is the first tree to start forming stands in land-uplift areas. It is relatively resistant to floods and fire. Animals do not readily eat its leaves and shoots, and it is relatively free of insect damage. The stem and roots are susceptible to rot fungi.

Scots pine (Pinus sylvestris) is an adaptable tree species with a wide distribution range. It is a pioneer species occupying treeless land, while at the same time it is a real or spurious climax species on nutrient-poor sand and gravel soils and on infertile peatlands.

Compared to other tree species, pine is the most competitive on nutrient-poor, water-pervious soils. It achieves its greatest size, with a tall straight trunk, on fresh glacial till soils. The competitive advantages of pine over broad-leaved species are its extended lifetime and large stem volume. On the most fertile sites and especially on soils containing limestone, the growth of pine is poorer than that of silver birch, aspen and spruce, and its trunk is of poor quality owing to the formation of thick branches.

Pine reproduces only through seeds. The formation of ripe seed takes two favourable summers, including the summer of flowering. In southern Finland there are 3 - 5 good seed years in every ten-year period, and one or two poor seed years. North of latitude 66°, the formation of ripe seed requires two successive warmer than average summers. Close to the timber line the interval between seed years can be as long as 10 years or more.

Most of the seeds are shed in the autumn. Seeds retained over the winter in the cones are shed in May-June. Most of the seeds fall to the ground within two lengths of the tree, the heaviest and best seeds very close to the tree. The seeds germinate and grow best on bare mineral soil and on soil with little humus and ground vegetation. Such sites include burnt ground, nutrient-poor sandy and gravely soils, and some types of peatland. Drought reduces the production of viable seedlings. On infertile sites in northern Finland the seeds lying on the ground can retain their germinating capacity for 1 - 2 years. Below old trees there are often seedlings of a wide age range awaiting the

Pubescent birch (Betula pubescens) is a pioneer species which invades those areas where other trees do not thrive. It has a meliorative effect on the soil. Pubescent birch grows well on oxygen-deficient soils, e.g. wetlands and their different drainage stages. However, its trunk rarely reaches the sort of quality and size required for plywood production. The photograph shows a ditched peatland near Oulu in northern Ostrobothnia. Pubescent birch-dominated forest accounts for 6% of the total forest land.

Aspen (Populus tremula) is a pioneer species which benefits from forest fires and warm sites, and it spreads vigorously through root suckers. Small groves of aspen occur in old forest. Aspen is of little importance for wood production because it is highly susceptible to rot, but it does play an important role in providing sites for rot fungi and insects. Aspen-dominated forest covers less than 1% of the forest land area. The photograph shows a climax forest growing on a Myrtillus forest site type in northern Häme.

Scots pine (Pinus sylvestris) is the most important tree species in Finland, and pine-dominated forests cover 65% of the forest land area. The best-quality stemwood grows in the esker districts of central Finland, the stemwood yield ranging from 3 to 8 m3/ha/year. The oldest pines grow in Lapland where they are 700 - 800 years old. The trunks of slowly dying, old trees turn silvery grey, which is highly appreciated for building purposes. The photograph was taken at Kiutaköngäs near Kuusamo in northern Finland.

Norway spruce (Picea abies) is the second most important tree species and it is dominant on 26% of the forest land area. The best-quality saw timber trunks come from southern Finland. On the most fertile sites the yield is about 10 m3/ha/year. The oldest trees are found in central Lapland where they have reached an age of 400 years. Spruce is easily destroyed by fire, and it has extended its natural distribution owing to effective fire control during this century.

Ash (Fraxinus excelsior) is one of the valuable broad-leaved trees, which also include oak, maple, lime and elm. It occurs naturally in small groups on fertile soil in southernmost Finland. The crown of ash is sensitive to freezing damage during the winter.

Lime or linden (Tilia cordata) regenerates in the shade of closed forests. As a result, lime occurs rather frequently as individual specimens or small groups. Pure lime stands are rare. The photograph was taken at Punkaharju in southern Savo.

opportunity to form a new stand. This ensures the regeneration of pine in a range of growing conditions. Seedlings which have been shaded by old trees for extended periods do not usually recover very well.

Pine demands light and growing space in all stages of its lifetime. Pine usually grows as approximately even-aged stands, apart from on the most infertile mineral soils and peatlands. If a stand is formed from seedlings and young plants of different age and size, self-thinning tends to convert it into an even-aged stand. The tree's self-thinning capacity is weaker on infertile and dry mineral soils, and dense pine stands often develop. Self pruning in the lower part of the trunk is better than that of spruce. Fully-stocked old pine stands form a hall of branch-free, tall pillars.

After the germling stage has passed the height growth of the young trees is rapid, reaching a maximum at an age of about 20 years on the best sites, 30 years on medium sites, and even later on infertile sites. Volume growth is at its greatest in young and middle-aged stands. Growth clearly starts to decrease at the age of 80 -120 years in southern Finland, and 120 - 150 years in northern Finland. Pine stands start to degenerate after 150 - 200 years. Individual trees can reach an age of 300 - 400 years.

The root system consists of superficial roots close to the ground surface, and deeper tap roots. The surface roots spread over a large area especially in coarse, water-permeable soil. Owing to its deep roots and strong trunk pine is rather resistant to strong winds. Pine produces only a superficial root system in the cold, severe climate of Lapland, and windthrow is a common phenomenon.

Insects can reduce the size of the seed crop. Birds and squirrels eat the seed, but they also spread them. The germlings are sensitive to drought and desiccating winds.

The seedlings are relatively resistant to frost. On land susceptible to ground frost, frost-heaving pushes the seedlings out of the ground and kills them.

Exceptionally cold, dry wind turns the needles brown during late winter. If there is only a thin snow cover or none at all, severe frost damages the fine roots close to the ground surface. In order to counteract nutrient disturbances, the trees transfer nutrients from the oldest needles to the younger ones, and then shed the old needles.

Many insect pests and fungal pathogens damage pine at all stages during its lifetime. They also make the seedlings less competitive compared to broad-leaved trees. The most serious fungal pathogen is pine branch twist (Melampsora pinitorqua), the intermediate host of which is aspen. Damage is more severe in planted than in naturally regenerated young stands owing to the lower number of stems/ha in plantations.

Pine is more resistant to stem and root rot fungi than spruce. Resistance to fire is also good. Windthrow occurs primarily in conditions where the root system is rather superficial owing to waterlogging or stoniness, or where trees which have become freed from a dense growing position have not had time to strengthen their roots.

The snow and ice which builds up in the crowns during winter can bend and break young trees growing in dense stands. This frozen snow and ice load is such a problem for pine in the hilly region in the eastern parts of northern Finland that spruce is the dominant species in these areas.

Moose, especially when their numbers are high, can cause severe damage in young plantations during the winter.

Pine is well adapted to forest fire ecology. The thick bark at the base of the trunk protects the tree against fire and heat. Ground fires do not reach the crowns, and a tree with a damaged trunk can still produce germinating seed, the seedlings growing well on the burnt ground. Pine is the dominant species on sorted sandy and gravely soils where broad-leaved species do not grow well and recurrent fires prevent the arrival of spruce. The presence of sandy and gravely soils, coupled with the recurrent occurrence of forest fires, is probably the main reason why pine is the dominant tree species at the northern timber line, although spruce also withstands the cold climate. The pines growing on infertile soils are examples of natural, almost single-species forests.

Norway spruce (Picea abies) is a clear climax species. It withstands shade, although not as well as Siberian fir (Abies sibirica), which is a species of the continental part of the boreal coniferous forest zone. The maritime climate in Finland has probably prevented its entry into Finland.

Spruce grows best and is the most competitive with respect to other tree species on the more fertile sites. It is also the natural dominating species on thin-peated spruce swamps.

Spruce starts to produce large seed crops at a later stage than pine. The seed mature in October-November during the flowering year. A considerable portion of them are shed very early on in spring when the wind carries them for long distances over the surface of the snow.

The interval between seed years lengthens to the north of latitude 62°. In the northernmost parts of the country ripe seed is produced only during warmer than average summers.

Insects and fungi destroy more spruce seed than pine seed. Birds and squirrels also eat them. On the average, the seed crop is very small five years out of ten, and large only once. During the rather rare good seed years the crop is extremely large and a lot of shade-tolerant seedlings are produced.

The seedlings remain small for the first 2 - 3 years, and even longer on shaded sites with mossy raw humus. Old seedling stands in actual fact often recover slowly when growing space becomes available. They are sensitive to frost, direct sunlight and drought. On fertile soil with good aeration and sufficient moisture the height growth of spruce seedlings is almost as great as that of pine seedlings.

Spruce can regenerate vegetatively. Branches touching the ground and covered with moss start to develop roots and eventually form a separate tree. In the archipelago and along the northern timber line groups of spruce have often been formed in this way.

Height growth in spruce stands reaches a maximum about 10 years later than that of pine stands, but growth continues for a longer period. The maximum dominant height in spruce stands on the best sites in southern Finland is 30 - 35 m, while for pine it is 25 - 30 m. The growth of spruce is poor on nutrient-poor raw humus sites; the trees remain short and stunted and their crowns reach down close to the ground surface.

Spruce stands start to degenerate earlier than pine, in southern Finland at an age of 100 - 130 years, and in northern Finland around 150 years. Individual trees can reach an age of about 400 years in southern Lapland.

The proportion of roots close to the soil surface is high, most of them being located in the humus layer and uppermost mineral soil. Vigorous tree growth requires good soil aeration. If there is standing water and a dense growth of mosses, spruce develops lateral roots above the plate roots, which improve its competitiveness. Spruce employs this strategy to maintain its dominant position on paludified soils in the north.

In cool humid conditions, the microclimate under spruce slows down the decomposition of plant litter and the proportion of nutrients bound in mosses, raw humus and humus increases. Site fertility is decreased. Under natural conditions only forest fire, which destroys the spruces and part of the humus and starts the succession of pioneer tree species, can restore the fertility and diversity of the ecosystem.

Bark beetles are the most serious enemies of spruce. They reproduce in trees which have lost their vitality. When the numbers of bark beetles increases sufficiently, they also move to healthy trees.

Stem and root rot fungi weaken spruce's competitiveness. They spread and weaken the stem and roots. Spruces with rot defects in the butt and roots are susceptible to wind damage. Windthrow also occurs on silt and clay soils in the autumn when the groundwater table is high and the soil waterlogged, and when the trees are growing in an exposed position.

Direct sunlight damages the bast layer under the bark of spruce stems and makes the trees susceptible to attack by bark beetles.

Owing to the shape of the crown and its growth mode, spruce is more resistant to snow damage than pine. Snow break promotes the spread of rot fungi. Spruce is better able to withstand short-term flooding than pine.

Owing to the thin bark and superficial root system spruce is sensitive to forest fire. Even a slight ground fire can kill spruce. The hanging crown reaching down to the ground and the presence of beard lichens in the branches promote the development of crown fires.

The natural regeneration of Siberian larch is not as good as that of domestic tree species in the Finnish semi-maritime climate. The germination percentage of the seed is low, below 20%. The photograph was taken in 1993 at Kitee Siberian larch stand.

FINLAND'S FIRST ARTIFICIALLY REGENERATED STAND

In 1842 the industrial pioneer, Nils Arppe (1803-1861) established a Siberian larch (Larix sibirica) stand with 4-year-old transplants and a European larch (L. decidua) stand by sowing at Kitee, about 80 km to the south of Joensuu in eastern Finland. The reason for planting the stands was to produce high-quality wood for shipbuilding, which was also the aim of Peter the Great in earlier times at Raivola on the Karelian Isthmus. The Tsarina of Russia, Anna Ivanovna, established the famous Siberian larch stand at Raivola in 1738.

Larches are not endogenous in Finland. The westernmost occurrence of Siberian larch is to the east of lake Onega in Russia, about 250 km to the east of the Finnish-Russian border. Siberian larch was an endogenous species in Finland before the last ice age. Remains of larch have been found in peatlands in central Lapland.

In 1932 the Finnish Forest Research Institute established sample plots in larch stands at Kitee. The most recent measurements were carried out in 1990:

PARAMETER	SIBERIAN LARCH *Larix sibirica*	EUROPEAN LARCH *Larix europea*
Age, yr	154	150
Stem number/ha	168	136
Basal area, m^2/ha	27	43
Volume, m^3/ha	409	584
Dominant height, m	35	37
Increment, m^3/ha (during the period 1982-1990)	4,8	4,9
Total yield, m^3/ha	722	850
Mean diameter, cm	46,5	65,4

The stem form of Siberian larch is better than that of European larch. The photograph shows a European larch stand at Kitee. The progenies of the Siberian larch stand at Raivola near St. Petersburg have a high-quality stem form.

Other tree species

Of the deciduous broad-leaved tree species growing in the temperate climatic zone, oak (Quercus robur) grows naturally along the south-western coast and in the archipelago. The conditions in the Åland islands more closely resemble the transitional stage between the boreal coniferous and deciduous broad-leaved forest zones. The distribution of ash (Fraxinus excelsior) is slightly wider. Other southern species are wych elm (Ulmus glabra) and European white elm (U. laevis), maple (Acer platanoides) and lime (Tilia cordata). Regeneration through sprouts has promoted the survival of lime during colder than average climatic periods.

Black alder (Alnus glutinosa) is a tree of the southern part of the boreal coniferous forest zone that grows on wet, fertile sites, especially floodlands and along the banks of watercourses. It occurs along the coasts on land rising up from the sea.

Yew (Taxus baccata) occurs naturally only in the Åland Islands, where it has a bushy form. The species spread to the Åland Islands during a warm period after the last ice age.

Curly-grained birch, which is a form of silver birch, grows naturally in the last slash-and-burn agriculture areas. Mountain birch (Betula tortuosa), which is a close relative of pubescent birch, is a bushy form growing in the forest tundra area.

A few dozen trial plantations of exotic species have been established in Finland. Siberian larch (Larix sibirica) is the most important from the point of view of timber production and as a landscape tree. Siberian fir (Abies sibirica) grows relatively well when planted. Lodgepole pine (Pinus contorta), which is a native of north America and grows fast when young, has been planted on about 2 000 ha. Its poor quality makes it unsuitable as saw timber, and it has no advantages over the domestic coniferous species. Beech (Fagus sylvatica) forms a closed forest canopy only in the Åland Islands.

Northern tree lines of the different tree species. Birches and aspen occur all over the country. Source: Lehto, 1969.

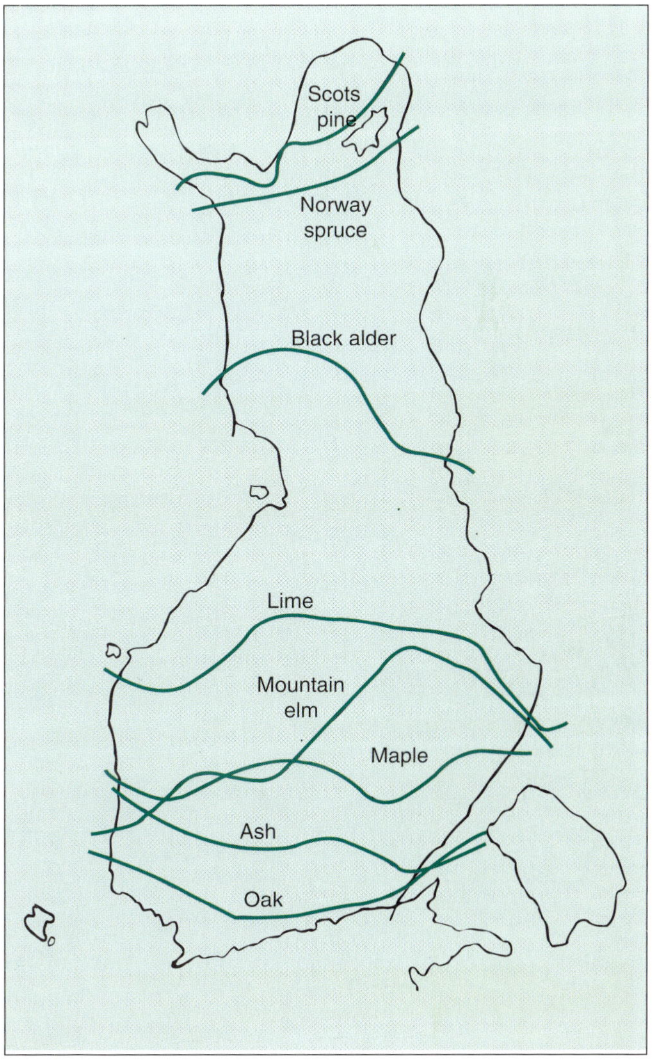

The dynamics of fire ecology

Forest fire is an integral part of fire ecology in the boreal coniferous forest zone. In natural conditions the forests are repeatedly renewed by the forest fires started by lightning, as well as to a lesser extent as a result of storm and insect damage. The pioneer trees and other plants invade the bare, burnt regeneration areas, and the succession of pioneer stands which eventually leads to climax stands, maintains great biological production and diversity.

It is difficult to recognise the role normally played by fire ecology in those parts of the northern coniferous forest zone, such as in Finland and the other Nordic countries, where effective forest fire control has prevented the outbreak of extensive forest fires, and where land and forest use have for long dominated the tree species dynamics. However, the original natural conditions are still to be found in the extensive unpopulated and sparsely populated areas of Canada and Asiatic Russia. In these countries the average area of forest burnt each year totals 2 - 3 million hectares.

The prevailing climate and occurrence of thunder storms determine the frequency of natural forest fires. According to studies carried out in Canada, the average frequency of thunder storms in July is 8 days near the southern prairies, and two days on the northern timber line.

A major forest fire requires warm, dry and windy weather, abundant inflammable ground vegetation and a closed tree stand that includes dry standing and fallen trees. A multistoried canopy structure also permits a ground fire to rise up to into the canopy to become a crown fire. Lichens and dwarf shrubs on the ground and beard lichens in the trees promote the spread of fire.

Fires recur more frequently and extensively on sandy and gravely soils. The higher the proportion of herbs and grasses in the ground vegetation and the finer the soil texture, the longer is the time interval between successive fires. Fires recur more frequently on the tops of hills and eskers and on southerly slopes than on shady, moist northerly slopes. The sandy terraces and slopes along rivers favour fires.

There is little reliable information about the occurrence of natural fires in Finland; throughout the period following the last ice age man first increased the frequency of fires and then, starting in the 19th century, effectively controlled them. Before the time of fire prevention they recurred at intervals of 50 - 100 years on sorted sandy and gravely soils, and at 200 - 300 year intervals on the more fertile and moister till and clay soils. The interval has been the longest in lush, wet groves, along the banks of watercourses and on water-logged soils. The intensity of major fires is well illustrated by the ash and charred wood that have been found in peat deposits.

In Finland's humid climate, natural forest fires were earlier usually only rather small ones. They left dead and damaged standing trees as well as isolated groups of seed-bearing trees, and a varied stand mosaic. During warm, dry, windy summers severe fires destroyed forest over thousands and tens of thousands of hectares.

According to historical records, forest fires during the 18th century were so extensive that they could be smelt in Stockholm, southern Sweden and even in North Germany. Despite fire-fighting efforts, major fires burnt tens of thousands of hectares during the 1930's. The most recent major forest fire destroyed 20 000 ha of forest at Tuntsa in eastern Lapland.

The biological diversity, i.e. the number of plant and animal species, is low immediately after a fire (stage 1 on page 83). However, there are species which are able to live in partly burnt trees and humus, but which are few in numbers or completely absent in the later stages of succession.

The longer the time which has elapsed since the last fire, the thicker is the layer of humus formed from partially decomposed plant litter, the greater is the amount of nutrients (especially nitrogen) bound in the humus layer in a form unavailable to plants, and the more acid is the humus. A forest fire releases these bound nutrients and reduces the acidity by 1-2 pH units. The solar energy reaching the ground in areas almost completely lacking a tree cover speeds up the decomposition of the plant litter and releases plant nutrients. Although a fire results in volatilisation losses of nitrogen from the soil, the amount of nitrogen in a form available to the plants is much greater after a fire (Stage 2).

The absence of a tree cover frees water and nutrients for other plants. Plants developing from seeds or by vegetative means rapidly form a multi-specied, luxuriant plant population, the high biodiversity of which is enhanced by the species living in burnt or rotten trees. There are abundant grasses, herbs and berry-bearing plants. Microbial activity is high, which speeds up the decomposition of the plant remains (cf. Chapter "Nutrient cycling").

Even minor fires usually kill spruce, which has thin bark and a shallow root system. Scots pine, protected by its thick layer of bark, retains its seed-bearing capacity. The butt of silver birch (Betula pendula) has a thick covering of bark which affords rather good protection against fire. Root and stump sprouts, and wind-borne seeds, speed up the role played by broad-leaved trees in colonising the bare ground.

Birches, alder (Alnus incana) and aspen (Populus tremula) are pioneer species proper. They colonise the moist and more fertile sites. Pine is a pioneer species on sorted, dry, thin soils. As succession proceeds the seedlings of the shade-tolerant spruce start to appear under the pioneer tree species. Spruce forms an understorey which slowly grows up into the canopy layer of the dominant trees, gradually capturing the growing space. If there are seed-bearing trees nearby and the conditions are favourable, spruce seedlings immediately appear on the burnt ground.

A short-lived alder stand changes within 30 - 40 years into a spruce-dominated stand. Spruce takes longer to overgrow birch and aspen stands, 50 - 100 years. The long-lived pine remains for a considerable time as an admixture in climax spruce stands (Stages 3 - 5).

Spruce is the climax tree species on all mineral soils, apart perhaps from the most infertile sandy soils. Pine is seemingly the climax species on sandy and gravel soils because the repeated occurrence of forest fires prevents the slower-growing spruce from getting a firmer foothold. The biological strength of spruce can be seen in the pine stands of modern times, which usually have an understorey of spruce owing to effective forest fire prevention.

As succession proceeds and the proportion of spruce increases, raw humus is formed on the surface of the soil. The humus binds many of the nutrients in a form unavailable to plants. The cycling of nutrients between the soil and the trees, and the biological production of the stand, decrease. The biodiversity of the stand is also reduced. The microclimate below old spruces is cool and damp. If the groundwater table lies close to the surface, Sphagnum mosses become established and paludification starts. As little as 1% of the total nitrogen reserves in the soil can be cycled between the trees and the ground.

The lack of plant-available nitrogen also impoverishes the species richness of the ground vegetation (Stage 6).

When the interval between successive forest fires is long, spruce-dominated climax stands start to degenerate. Dead trees leave gaps in the stand. The ground vegetation flourishes, broad-leaved trees develop from sprouts, and spruce seedlings appear. Trees whose butt and roots are weakened by rot are felled individually and in groups by the wind. Bark beetles breed in trees with weakened vitality as well as those felled by the wind. They also spread to the trees surrounding windfall openings and enlarge the regeneration area. When the opening is sufficiently large, a broadleaf-dominated stand and spruce undergrowth are formed. Regeneration which occurs in this fashion is called a minor succession cycle, as opposed to the major cycle which starts from the bare ground.

Severe storms fell trees of all ages, often over large unbroken areas. In addition to forest fire, storms and insects help to maintain the cycle of pioneer and broad-leaved species.

Insects which feed on the needles of conifers play only a minor role in forest regeneration under Finnish conditions. Under favourable climatic conditions, populations of sawfly (Diprion spp.) species can become so large that they kill all the pines over extensive areas. In America, on the other hand, insects play a considerable role in regenerating the forest. Insects which feed on buds and needles destroy thousands of hectares of spruce and fir forest, reduce the "over-large" tree stand and regenerate the forest. Insects and spruces are said to live together in a form of symbiosis in which the regeneration cycle of the stand lasts about 60 years.

Coniferous forests in a natural state also contain plant populations growing on sites which are never or only very rarely affected by forest fire. Such sites, i.e. fire refuges, include floodlands along riverbanks, peatlands and fertile moist groves. They form a sanctuary for those plants and animals which have become adapted to conditions not characterised by repeated forest fire. The banks of water-courses and river valleys form the routes along which many species spread into new areas. Examples of this are the upper reaches of rivers flowing from Finland into Russia which contain many eastern species.

Herb-rich forests usually cover only a small area. The soil contains abundant calcium. The surface soil is moist, and there is usually flowing water apart from during the driest spells. Herbs, grasses and large ferns dominate the ground vegetation. The flowering plants include species unique to this type of site. Easily ignitable dwarf shrubs and mosses are few in number or completely absent. The diversity of the vegetation is demonstrated by the fact that the different species flower at different times during the growing season. The flowering plants also have a longer growing season than those found on other types of site.

Recurring forest fires affect the occurrence frequency of plant and animal species and possibly also promote specialised development. An excellent example of this is the North American lodgepole pine (Pinus contorta), which is one of the tree species best adapted to fire ecology. Rapid regeneration following a fire is based on the fact that the cones are retained in the tree crown for many years. They are closed but contain germinatable seed (serotinous cones). The cones open in the heat of a fire, and the seeds fall to the ground and germinate immediately after the fire. The lodgepole pine provenances which grow on land where fires most frequently recur have the most tightly closed cones.

The Arctic bramble (Rubus arcticus), which grows in

Finland, has developed its own adaptive strategy in fire ecology. It is a dioecious plant species. i.e. individual plants produce only male or female flowers. Berry production requires cross-pollination in the sort of stands, containing both male and female plants, that occur in repeatedly burnt areas. When fires are prevented the Arctic bramble occurs in vegetatively reproducing stands containing only female or male flowers and hence there are no berries.

The narrow watercourses normally to be found in forested areas play an important role in nutrient cycling and the whole spectrum of diversity. The leaching of nutrients into the watercourses increases immediately after a fire. The aquatic vegetation and that growing along the banks flourishes. Herbivores and predators increase. Some of the leached nutrients are returned to the forest in their debris and excreta. Nutrient cycling and diversity increase.

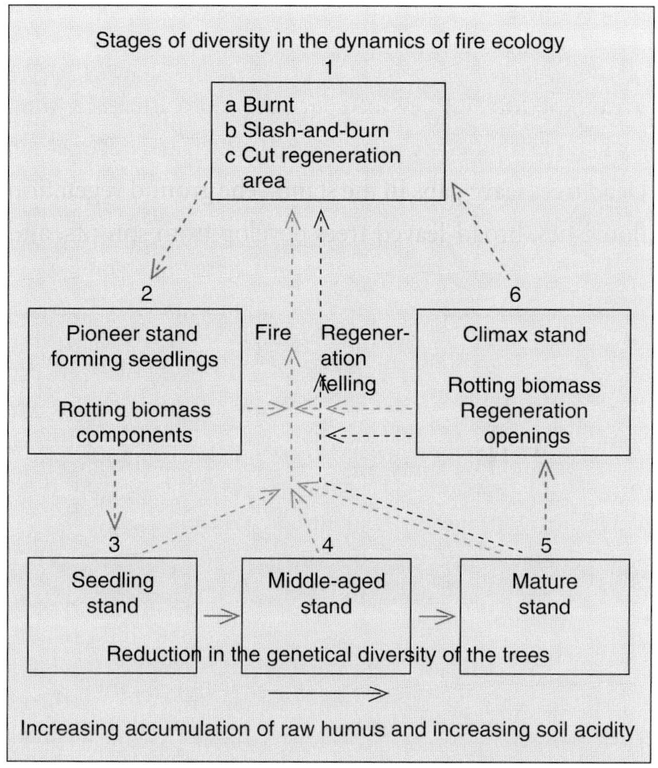

The major succession cycle of a renewable forest in fire ecology. Fire and other damaging agents can initiate regeneration at any stage. Regeneration of degenerated climax stands takes place through the minor succession cycle in the openings left by dead trees. In commercial forests mature stands are regenerated.

Examples of strategies used by plants to survive fire

Thick butt bark protects the trunk against fire and the seed-producing capacity of the tree is retained, Scots pine.

Fire-susceptible conifer, but the crown contains many serotinous cones that are opened by the heat of the fire, shedding ripe seed onto the ground, lodgepole pine (Pinus contorta).

Tree produces abudant, air-borne seed, birch.

Stems of the plant grow in dense, moisture-retaining clumps capable of forming new shoots, hair-grass (Deschampsia flexuosa).

Sprouting parts are so deep in the ground that they are not killed by fire, bracken (Pteridium aquilinum).

The seed remain viable for long periods and germinate following the heat of a fire, (Geranium bohemicum).

Area of forest fires and prescribed burns

1950 –54	1955 –59	1960 –64	1965 –69	1970 –74	1975 –79	1980 –84	1985 –89	1990 –93
Forest fires 1000 ha/y								
3.5	3.5	5.4	1.5	1.3	0.6	0.4	0.3	0.6
Prescribed burns 1000 ha/y								
10	31	19	7	1	1	1	3	2

If the lightning strike is only a light one, the bark becomes detached from the pine trunk and a stripe is left down the trunk. This "candle of big lumberjack" has been left to grow."

LIGHTNING AND FOREST FIRES

Rapid changes in the weather cause thunderstorms, and during dry summer weather lightning readily starts forest fires in coniferous forest. In the south-western corner of Finland thunderstorms occur on about 20 days a year, and in Lapland 5. During the average summer in Finland lightning strikes over 50.000 times, and on the average there are three heavy thunderstorms every year. The frequency of forest fires started by lightning in the southern parts of Fennoscandia is much greater than that in the north.
Lightning often strikes tall trees, the type of damage usually being characteristic of the different tree species. The damage caused by lightning varies from a narrow stripe down the bark of the trunk to explosive destruction of the whole trunk. Lightning can also kill all the trees in an area covering a few hundred square meters. The signs of this type of damage do not appear until autumn the following year when the trees die from the crown downwards. Lightning damage of this kind has been found on hill slopes.

This silver birch has been struck a smashing blow by lightning. Large pieces of wood were found up to 50 m away.

All countries have legislation and fire-fighting organisations to prevent forest fires. Aerial fire spotting is effective, and alarm systems based on satellite image interpretation are being developed. Man is retarding the natural regeneration of the boreal coniferous forest.

The extent of insect damage can be as great as that of forest fires in the boreal coniferous forests. Mass outbreaks of pine sawflies destroy extensive areas of pine forest if the damage recurs for a number of years in succession. The photograph was taken at Lauhanvuori, northern Satakunta. It is easy to control the damage in commercial forests, but the insects are allowed to multiply in national parks. A new forest returns slowly to the area; the patches of mineral soil exposed when the dead trees fall over become covered with new seedlings.

The larvae of pine sawflies (Diprion pini) eat the current and previous years' needles of pine during July-August. This species is the most damaging sawfly because it consumes all the needle biomass. Repeated outbreaks kill pine stands.

THE PIONEER STAGE IN CONIFEROUS FOREST SUCCESSION AT EVO

The absence of proper forest fires has resulted in the use of burning in the state commercial forests and national parks since the beginning of the 1990's. This has been done as part of research into coniferous forest ecosystems, and to develop forest regeneration techniques for the national parks. The Tuohimetsä forest area at Evo, near Lammi in southern Finland, was burnt for research purposes on the 1st of June 1992. The area was about 9 ha large and located at an elevation of 160 m a.s.l. The site was of moderate fertility (Myrtillus forest site type).

The aim of the burn was to replicate a forest fire as closely as possible, and then to follow the early stages of coniferous forest succession by means of plant species inventories. The soil in the depressions in the area is rather fertile and on the hillocks stony. The tree stand in the area to be burnt was felled in winter 1991/92, leaving a number of larger pines in the edge belt as seed trees, and the central part of about 1.5 ha completely uncut. The stand was a naturally regenerated one that had been thinned on a number of occasions. The stand included 160-year-old pines and 80 to 120-year-old spruces, the average volume of the stand being about 300 m^3/ha before cutting. There were also a few silver birches, pubescent birches and aspens in the area.

168 m^3/ha of the more valuable, large spruces in the central area were felled, and the logging residues promoted the spread of the fire up into the tree crowns. The volume of the standing trees in the central area at the time of the burn was 150 m^3/ha, and there was a dense understorey. The number of trees less than 1.3 m high was 8 000/ha, half of which were spruce. When the fire was started the weather was partly cloudy and rather favourable; the wind speed varied from 3 - 6 m/s, and the logging residues and ground vegetation were half dry. The logging residues in the edge belts promoted the fire and the slopes helped to lift the fire up into the tree crowns.

The old spruce forest had a thick moss cover and was growing on a stony site of the Myrtillus forest site type.

Many leguminous species are pioneers which colonise the exposed, burnt ground after fires and add nitrogen to the soil. Sea pea (Lathyrus japonicus) is the first plant to colonise the sandy soil rising up out of the sea. The photograph shows wood vetch (Vicia sylvatica) on burnt ground. The increase in pH caused by the fire promotes the activity of nitrogen-fixing bacteria. The competitiveness of legumes is improved by the severe fires that substantially reduce the total nitrogen reserves.

The researchers group Henrik Lindberg, Raili Suominen, Tiina Tonteri, Eevastiina Tuittila, Seppo Tuominen, Ilkka Vanha-Majamaa and Ari Väänänen has studied the pioneer flora after five. Species benefited by the burning were Vicia sp., Veronica officinalis, Calluna vulgaris, Ceratodon purpureus, Potytrichum juniperinum and Marchantia polymorpha.

The spread of aspen is accelerated by forest fires. The rhizomes and roots of aspen produce fast-growing sprouts which, in this case, have formed a plant community 20 - 30 m away from the dead mother tree within two months after the fire.

The burnt ground is covered in vegetation within two growing periods. Abundant species include fireweed (Epilobium angustifolium), hair-grass (Deschampsia flexuosa) and small reed (Calamagrostis arundinacea). However, moose (Alces alces) have browsed on the aspen sprouts on a number of occasions during the first growing season.

Researcher Raili Suominen is inventorying the vegetation cover and number of seedlings on the sample plots. Within the first year the central area contained 22 aspen seedlings/m^2, 2.5 spruce germlings/m^2, 2 pine germlings/m^2 and 8 rowan seedlings/m^2. During the second year the density of the aspen seedlings decreased by 50%.

The fire was started after a one-week sunny period but neither the forest nor the humus layer had time to dry out sufficiently. The fire mainly proceeded as a ground fire. On Lahnasuo island, pine accounted for 37% of the basal area of the stand, spruce 45% and birch 18%.

THE PIONEER STAGE OF CONIFEROUS FOREST IN THE PATVINSUO NATIONAL PARK

Two 1-ha pockets of upland forest, isolated by the surrounding peatland, were burnt in the Patvinsuo National Park at the end of June 1989. Patvinsuo lies at an elevation of 170 m a.s.l. in Eastern Finland, about 70 km to the north east of Joensuu. The purpose of the burn was to regenerate the forest and to investigate the different stages of floral and faunal succession characteristic of fire-based ecosystems. The forest on both islands consisted of a mosaic of the Myrtillus and Vaccinium forest site types, interspersed by small rocky outcrops.
The stand on Lahnasuo island was in a natural state and dominated by Scots pine with a mean age of 90 years. Some of the pines were over 200 years old, and there were also some dead, standing trees. A spruce undergrowth occurred sporadically in hollows and there were a few large birches. The volume of the standing trees was 130 m^3/ha, and that of dead fallen trees 15 m^3/ha.
Parts of the pine stand on Surkansuo island were two-storied: a naturally regenerated dense stand about 9 m high, with a sparse overgrowth of pine. There were some old birches mixed in amongst the pines and a few clumps of small grey alder *(Alnus incana)*.
The burn was preceded by a week-long sunny period, and the fire at both sites primarily progressed in the form of a ground fire. The spruce undergrowth and beard lichens on the trunks and branches of the old trees enabled the fire to climb up into the crowns of isolated spruces on Lahnasuo island. The fire continued to smoulder in the butts of a number of old, dead standing pines, which eventually fell down.
Changes in the tree stand, ground vegetation, conk fungi *(Polypores)*, beetles, soil molluscs, soil fauna, soil nutrients and microbes have been followed in the study. The burn was not very intense; the acidity of the soil was reduced by only 0.4 pH units. In the strong fires used in prescribed burning the reduction in acidity can be as much as 2 pH units. The calcium and magnesium contents in the soil doubled, and remained at a higher level for some time. The potassium content increased by 50% immediately after the fire, but fell to close to the starting level during the year following the fire. The mycorrhizal associations of the trees were drastically reduced by the fire.
According to the beetle survey, there are 520 species in the area. The ground beetles, *Agonum quadripunctatum*, *Stephanopachys substriatus* and *Acmaeops marginata*, for instance, have benefited from the fire.

Pine, spruce and birch seedlings have reached a height of 10 - 50 cm within four years after the fire. Burnt and dead trees have fallen after their roots have rotted. The ground is not yet completely covered by seedlings. The photograph is of Lahnasuo island.

The ground is covered in ash after the fire. The large pines have survived, but the spruces, birches and small pine have died on Lahnasuo island.

The release of nutrients through soil weathering and the deposition of nitrogen from the atmosphere compensate for the nutrients lost from the ecosystem. When forest fires are prevented, the fertility and biodiversity of the watercourse and bank biotopes decrease in comparison to the situation in areas subjected to recurrent forest fire. The sequence of events characteristic of fire ecology maintains the nutrient cycling and great biological production and diversity of natural forests. The overall spectrum of diversity consists of treeless, burnt ground, plant succession running from pioneer species to climax stands, stand mosaics and plant communities not subjected to fire. The plant, animal and microbial species follow the individual stages of succession over time and space.

The above-ground parts of this grey alder (Alnus incana) were killed in the fire on Surkansuo island and there were no signs of life one year after the fire.

However, the grey alder was not completely dead because it slowly produced new sprouts from its roots. Five years have elapsed since the fire on Surkansuo island.

AFTER THE KITSI FOREST FIRE IN 1992

The forest at Kitsi in northern Karelia was burnt at the beginning of June 1992, when a strong northerly wind (20 m/s) spread the fire from a nearby prescribed burn. The area lies at an elevation of 180 m a.s.l. and is located at a distance of about 40 km to the east of Lieksa. The fire was preceded by a warm period and 143 ha of forest were burnt. A crown fire immediately destroyed the unthinned spruce-dominated forest, but remained as a ground fire in the old managed pine stand that had been intensively dry pruned. The site in the spruce-dominated forest is of the Myrtillus forest site type, and in the pine forest of the Vaccinium forest site type. Most of the burnt area has been left as a field experiment for fire ecology studies.

*Life immediately returns to the burnt forest after the last smouldering remains of the fire have died out. The resin from damaged and dead trees has attracted insects, the most common of which are the pine weevil (*Hylobius abietis*) and the sawyer beetle (*Monchamus sutor*) shown in the photograph.*

*One year has passed since the fire in the commercial mature pine stand. The fire proceeded as a ground fire and completely destroyed the understorey stand. Blueberry (*Vaccinium myrtillus*) and lingonberry (*V. vitis-idaea*) in the field layer are recovering, but not crowberry (*Empetrum nigrum*). The site is of the Empetrum-Vaccinium forest site type.*

The stems of small, thin-barked willow species are destroyed in forest fires, but the rhizomes have recovered within one year to form flower and seed-bearing shoots.

Fire destroys the anthills of the red ant (Formica rufa) and kills most of the ants. However, the species returns rapidly to its devastated home, and there is no shortage of needles for the building work.

When trees are killed by forest fire their symbiotic mycorrhizal fungi decline or disappear, while the numbers of rot fungi increase in the soil. The fungal flora is usually impoverished by fire. Pine fire fungus (Rhizina undulata) appears in large numbers in the area within two years after the fire owing to the increased nutrient availability. Another common species is shown in the photograph - Geopyxis carbonaria which occurs in the most fertile parts of the hollows.

THE BURNT FOREST AND WETLAND AREA AT OSKARSHAMN IN SWEDEN

In the middle of August 1983, 650 ha of mainly scrubby forest were burnt on the western side of the town of Oskarshamn in southern Sweden. A 65 ha fire reserve has been left in a natural state in the Hammarsebo area for research purposes. The old pines survived the fire. About one year after the fire there were 90 pine and 30 spruce germlings per square meter on the intensively burnt ground. There were 5 silver birch, 15 pubescent birch, 60 aspen and 7 goat willows per square meter. Game animals and especially hares have intensively browsed the broad-leaved trees. Five years after the fire 60% of the conifers and 50% of the birches and 20% of the aspens had survived. The 2 cm-thick, unburned humus layer has proved to be a poor substrate for the germlings.

Spruce has developed naturally below the old pine stand growing on this rather dry site. In sunny places the lower spruce branches extend down to the ground surface. If fire breaks out in a forest of this sort during a dry period it will burn easily and the fire spread rapidly up into the crowns.

A similar forest to the one in the previous picture but 10 years after the Hammarsebo forest fire. All the spruces are dead and have fallen to the ground, but some of the pines have survived. Large numbers of birch, pine and spruce seedlings have appeared on the site.

The fire was so intense that it spread to a pine swamp. Most of the trees were killed, but the surviving pines and fringing forest have seeded the ash-fertilised swamp. The effect of ash fertilization can be seen in the bushy crowns of the surviving pines.

The dwarf shrubs, grasses and herbs, e.g. the spotted orchid (Dactylorhiza maculata), growing on the pine swamp have become very lush as a result of the fire.

27 YEARS HAVE PASSED SINCE THE REIVO FIRE IN NORTHERN SWEDEN

Reivo is located near Arvidsjaur in northern Sweden. The conservation area covers almost 10 000 ha. Most of the pine-dominated forest is over 250-300 years old. Pine grows on the warm southern and western slopes, whilst spruce is to be found on the moist, cold northern slopes. There are few broad-leaved trees in the area. The conservation area is located in the reindeer-husbandry region.

A total of 300 ha of over 200-year-old coniferous forest were destroyed in the Reivo fire in the middle of June, 1966. The forest lies at an elevation of about 400-500 m a.s.l. The fire spread into the conservation area from a prescribed burn carried out in the neighbouring commercial forests. The weather preceding the fire was dry, and the crown fire that developed almost completely destroyed the tree stand. The forest in the conservation area has been allowed to develop undisturbed, but the commercial forest was regenerated by cultivation.

The forest fire resulted in an almost completely treeless area extending over more than 200 ha. This has considerably slowed down natural reforestation. Dead, standing trees remain unaffected by rot for several decades and burnt pine stumps for as long as 200 years.

A large number of conifers were killed by the fire in the area. The seedlings are now reaching the snow limit 27 years after the fire. However, there is a well-developed, broad-leaved seedling stand on the bluff in the background.

Reindeer and moose eat the pioneer broad-leaved trees, thus retarding reforestation in the area. The birches, aspens and willows have reached a height of 2 - 6 m on the inaccessible bluff.

The pine stand cultivated in the burnt area in 1966 has reached a height of 3-6 m within 25 years. The dead, standing trees left in the burnt area can be seen behind the seedling stand.

THE GREAT TUNTSA FIRE OF 1960

The Tuntsa wilderness area is located in the northern part of the district of Salla, 120 km to the north of the Arctic Circle, and is bounded on the east by the Finnish-Russian border. The closest settlement is the mining town of Kovdor in Russia. Tuntsa belongs to the protection forest zone, designated by law in 1922, in which commercial forestry is practised with extreme care under the supervision of the Finnish Forest and Park Service. The lowest valleys lie at an altitude of 260 m a.s.l., and the highest treeless fells at 630 m a.s.l.

Summer 1960 was one of the hottest in Lapland this century, and even the wetlands dried up. Lightning ignited the bone-dry ground at the end of June, and the forests on the Russian side of the border also went up in flames. The fire developed into a major fire, and on the Finnish side 500 local men and conscripts were engaged in the fire-fighting work. The present director general of the Finnish Forest Research Institute, Prof. Eljas Pohtila, participated in the fire-fighting as a 16-year-old schoolboy. The fire continued to rage in the area for one month. Extinguishing the fire was difficult because there were no roads in the area at that time, and the fire-fighting equipment was rather rudimentary.

The burnt area on the Finnish side totalled 20 000 ha, of which 10 000 ha were forest. The fire also burnt 100 000 ha of forest in Russia. The spruce-dominated forest was destroyed, apart from stands located along the riverbanks. Although the small trees in the pine stands were killed, most of the large trees survived owing to the thick layer of bark at their butts.

Damage to the tree stand was not very great. 300 000 m^3 of timber were felled in the area. However, the reindeer lichen pastures were destroyed for decades. Nowadays there is a rather large reindeer population at Tuntsa, especially now that the summer pastures have recovered. Moose, large predators and eagles live in the area.

A start was made on reforesting the area during the 1960's. The soil was prepared by ploughing, harrowing or scalping, and pine was sown and planted. The plantations covered 9 000 ha. Most of the plantations have since failed owing to the extreme climatic conditions and the use of unsuitable seed origins. A good pine seed crop was obtained at Tuntsa in 1973, and the transplants grown from this seed have been very successful. Natural regeneration is also slowly taking place in the area. Natural reforestation along the timber line at Tuntsa shows that even a major forest fire cannot prevent the return of coniferous forest species. The burnt areas form a mosaic regulated by the topography; small fire refuges are always to be found in the marginal zones along the watercourses.

The inner page on the front cover of the book shows a view of the spruce area, burnt 33 years ago, which is now covered in a dense young birch seedling stand. Bud break in birch occurs around midsummer, and the last snow disappears from the ravines in the fells by the middle of July. Spruce has survived along the stream valleys, its seed subsequently spreading to the neighbouring slopes. The trees produce seed only during exceptionally warm years.

The inner page on the back cover of the book shows an old, thick-mossed, climax spruce stand that is not capable of regenerating naturally at all. The birches in the spruce stand are dying, partly from old age and partly as a result of the extensive damage caused by the autumnal moth (*Epirrita autumnata*) at Tuntsa in 1965-66. Birches which are defoliated in the first half of summer and subsequently have time to produce new foliage during the growing season, survive the damage caused by these larvae.

A few isolated birches survived the light fire. The butts are scarred, but rot fungi cannot easily penetrate the trunks. The fire occurred 30 years ago.

Isolated pines have produced an even-aged, young stand. The year when the seed year occurred can be obtained by counting the number of annual shoot segments on the seedlings.

The pine stand growing on the Uhkunaho upland site in Suomussalmi, Kainuu, was burnt 42 years before this photograph was taken. The fire was a light ground fire and the butts of the pines were not damaged, but the spruces died. The surviving, dense old pine stand has produced a new generation of trees in the openings. The seedlings are slender with thin branches owing to the shady conditions.

Most of the pines at Uhkunaho were not burnt and a 3-6 m high spruce stand has developed there. The spruce seedlings were rather small at the time the fire burnt through the neighbouring stand. The general decline in forest fires has promoted the development of spruce undergrowth.

The forests along the banks of the Oulankajoki River in the Oulanka National Park in northern Finland were burnt during the 1890's. A pine stand has regenerated after the fire, but the dense stand has since been affected by many types of damage. Spruce has appeared on the dry upland sites, and will eventually dominate the infertile sites on sorted soil unless fire again interrupts this progression.

THE LARGE FOREST FIRE AT KIHNIÖ-PARKANO IN SOUTHERN FINLAND IN 1933

Summer 1933 was exceptionally dry and warm in western Finland. A number of forest fires broke out in the western and central parts of the country in July, and in the largest one 5 500 ha of forest were burnt. A total of 31 000 ha of forest were destroyed in 300 fires in Finland in 1933.

It took three years to fell the trees on the burnt area. Slash-and-burn agriculture was still a familiar practice to the farmers in those days and rye was sown on the most fertile sites fertilized by the wood ash.

A mixed pine-birch stand, growing on a site of the Myrtillus forest site type, has developed in the area. The stand has been growing in a natural state for 57 years since the fire, apart from some light, young stand tending. The canopy has closed and some of the trees have been killed off by competition.

A comparison stand, managed as commercial forest, that regenerated at the same time on a nearby site. The young stand has been tended and thinned once. The volume of the stand is about 200 m^3/ha and the estimated increment 8 m^3/ha/year.

They continued to slash-and-burn for a few years, but the last patches of rye were also sown with pine and spruce seed. Pine was sown on the infertile burnt uplands, and spruce on the fertile sites. Birch seeded naturally on the burnt ground. The state supported reforestation with forest improvement funds.

The old aspens killed in the fire are popular nesting sites. Many woodpeckers nest in the rotten, hollow trees during the spring, the common swift (Apus apus) being the last bird to nest there during the summer.

In the evening the great spotted woodpecker (Dendrocopus major) searches among the same rotten holes for a suitable place to spend the night.

Geranicum bohemicum is one of the plants that has adapted to fires and grows on the most fertile sites. The seeds start to germinate when a forest fire raises the soil temperature above 50oC. The plants in the photograph were flowering on a herb-rich site at Lohja, southern Finland, in 1990. The tree stand was felled in 1988, and the seeds lying dormant in the soil started to germinate after harrowing in 1989. The species has become rather rare during the past few decades.

THE "MINOR REGENERATION CYCLE" CAUSED BY STORM AND SNOW DAMAGE IN THE PISAVAARA NATURE PARK

Wind and snow eventually form openings as the tree stand degenerates in ageing coniferous forest. The conditions are suitable for forest regeneration when the crown cover is removed or thinned naturally. Seedlings slowly appear on the uprooted butts of fallen trees, on the exposed mineral soil and along the edges of rotting trunks. The phenomenon is called the "minor regeneration cycle", while the dynamics of fire ecology are called the "major regeneration cycle".
Pisavaara is located 40 km to the south-west of Rovaniemi and 30 km to the south of the Arctic Circle. The highest point on the tree-covered hill, Liljalaki, rises to an altitude of 282 m a.s.l. The upper slopes are cloaked in fertile spruce stands, and the lower, less fertile slopes mainly with pine forest. The 5 000 ha area has been maintained as a pristine nature park since 1938. A severe gale felled a large number of trees in the park in autumn 1982. Extensive snow damage has occurred repeatedly on the tops of the hill.

The storm in autumn 1982 produced a 1 ha opening at the bottom of the hill slope; the site is of the herb rich forest type with a covering of spruce. Some of the spruce seedlings in the understorey have recovered within 10 years, but a moss cover is still preventing forest regeneration.

Forest regeneration has started in the exposed mineral soil around the uprooted butts of large fallen trees. 10 years have passed since the storm.

Snow breaks the trees in the hilltop spruce stands almost every year. Stand regeneration starts on the rotten trunks and rocky hummocks. Canopy closure is never reached owing to recurrent snow damage and the humid climate. High humidity retards regeneration of the pioneer tree species and improves that of the climax species.

THE LAHOSAAJO FIRE REFUGE

Forests which have been allowed to grow undisturbed for centuries are rare. If the forest is surrounded by watercourses or peatland, it is less likely to burn than unbroken forested areas. It has been estimated that the forest island of Lahosaajo, isolated by the surrounding peatland, has remained untouched by fire for at least 1 000 years. No charcoal residues have been found in the soil. The site is located at Teuravuoma, near Kolari in western Lapland, 80 km to the north of the Arctic Circle.

Owing to their rarity, fire refuges are interesting as climax forests and also because they contain a group of species adapted to highly specialised conditions. Because fire has been absent for hundreds of years, the stand is dominated by spruce. The largest spruces are 300 years old. The diameter at breast height of the largest spruce is 43 cm, and height 23 m. The dominant height of the stand is 15 m. The volume of the stand is 55 m^3/ha, and two thirds of it is spruce and one third pubescent birch. The annual increment is clearly less than 1 m^3/ha.

Pubescent birches grow sparsely and do not reach the same dimensions as spruce. The birches cannot reproduce sexually because the soil is covered with a thick layer of moss, but they do survive on the site by reproducing through stump sprouts. The crown canopy has not closed owing to the deterioration in site quality, and growth is also weakened by the accumulation of humus. The amount of plant-available nutrients is low. The regeneration capacity of the spruces is also poor, and spruce seedlings are mainly restricted to the butts of fallen trees and the edges of large, decomposed anthills.

The forest contains a lot of rotten trees felled by the wind and snow. Rot fungi and many species of beetle live on and in them. Researcher Juha Siitonen has so far found 232 beetle species living in rotten trees, many of them rare. The number of species is large considering the conditions at this site. This is probably due to the undisturbed state of the stand, which has continued for so long that all the species which require this sort of environment have had time to colonise the site. The regeneration cycle is continuous and the spruce stand always contains trees at different stages in the rot process.

Species which are adapted to fire refuges represent only a few per cent of all the beetle species discovered so far. They include Agonum mannerheimi, Pytho abieticola and Monochamus ururovii. Beetle species which live on rotten trees in fire refuges (so-called saprochylites) require humid conditions and a shady microclimate, and they cannot spread easily. In these respects the refuge species differ rather radically from the pioneer species associated with forest fires. It is important to maintain the continuity of the environment for these refuge species.

The crown canopy of the old Lahosaajo climax spruce stand does not close owing to the declining biological activity of the site. As an ecosystem it is a rare key biotope in the boreal coniferous forest zone. A group of characteristic species live in fire refuges.

The butts of the large spruces are rotten and the trees are easily felled by storms. The continuous supply of rotten stems maintains a beetle population adapted to fire refuges.

The old forest contains trees in different stages of decay that support a diverse population of conks (Aphyllophorales) and beetles. One common conk species is Trichaptum abietinum.

MAN'S FOREST

The forest and the people who settled in Finland accepted each other. Without the forest, its game, trees and the solar energy stored in the trees, man would have not been able to develop an independent agrarian, and subsequently affluent industrialised state. The forest has been a shield against the severe climate. Its bounty has helped the population survive when the crops failed. The forest has also protected people against the tribulations of war

Hunting, fishing and nomadism

The first people, the Lapps (Same), arrived in Finland over 9 000 years ago hard on the heels of the retreating ice sheet. They migrated from the east and south-east as fishermen and hunters along the shores of the White Sea and Arctic Ocean, some of them settled in the fell lands of the north and some in southern Finland. The latter group were pushed up into the north by the Finns, and become the forest Lapps of Lapland. The Finns arrived from the south-east and south. Germanic tribes also came in from the west.

The first inhabitants were fishermen and hunters. Moose and beaver were the most important game. The edible parts of wild plants provided additional food. Wood was used to build dwellings, tools and other essential equipment. It provided heat for dwellings and for preparing food. Above the northern timberline housing was made from peat, which was also used for heating. Peat dugouts were still being used as dwellings at the beginning of this century.

Wooden boats and ships enabled people to settle in the archipelagos along the coast and to move inland up the watercourses. Ships opened up trading routes to far away lands.

Semi-domesticated reindeer provided the Lapps with a means of sustenance and a source of wealth. Seasonal variation in the location of suitable pastures and the availability of lichens, herbs, grasses, sedges and edible fungi meant that the Lapps had a nomadic lifestyle. Fishing grounds lay along the trapping routes. The trapping of wild reindeer encouraged the movement of the communities.

Valuable furs attracted hunters and merchants to the forests of the north. The soldiers who followed them tried to expand the domains of their own countries. Possessing power meant being able to levy taxes. Subservience was paid for with furs, meat and fish.

Hunting, fishing and nomadism changed the natural forest. The effect was greatest on the flora and fauna surrounding the permanent settlements. The effects of reindeer husbandry were restricted to the north and the timber line on the fells. Grazing changed the relationships between the ground vegetation and the trees. Fires were lit in order to improve the quality of the pastures. Spruces festooned with horsehair lichens were felled to provide feed for the reindeers during the winter.

The greatest changes in the tree species composition occurred to the north of the closed forest region, in the valleys of the rivers flowing into the Arctic Ocean and along the coasts. Conifers were used for buildings, boats and as firewood. The only signs of pine in extensive parts of the far north are the place names.

The closed forest was too strong for man alone to bring change, or to lower the northern timberline. But when cold periods prevented the formation of germinating seed, man lowered the timberline with the help of the

REINDEER IN THE BOREAL CONIFEROUS FOREST

The doubling in the reindeer stock during the past 20 years has had a detrimental effect on the winter pastures. The size of the winter population is at least 200 000 reindeer. Supplementary feeding has to be provided during the winter, and it is nowadays a common practice throughout the whole reindeer husbandry region. The natural winter food of the reindeer are reindeer lichens, hair grass (Deschampsia flexuosa) and the beard lichens (Usnea spp.) and horsehair lichens (Alectoria spp.) growing on tree trunks. When the snow cover is thick and frozen in late winter, cratering for lichens is difficult and beard and horsehair lichens represent almost the only source of food. Forestry is criticised for felling and regenerating the beard-lichen forests and pine stands on the lichen heaths. Birch plantations cannot be established in the reindeer husbandry region without fencing. During the summer the reindeer browse on the shoots of young, naturally regenerated broad-leaf stands and damage and retard the regeneration and growth of birch stands.

Reindeer are allowed to graze freely, independent of the land owner, in the reindeer husbandry region. The animals are rounded up around mid-summer for the calves to be marked, and in the autumn for the annual slaughter. There are about 7 000 reindeer owners, the reindeer husbandry region being divided into 57 reindeer husbandry districts.

Winter feeding is common throughout the reindeer husbandry region. Supplementary feeding has to be resorted to when the reindeer stocks exceed the carrying capacity of the winter pastures.

Reindeer are also allowed to graze freely in the national and nature parks. Grazing causes permanent disturbance in the ecosystems; the lichen stands do not recover, herbs and grasses are browsed, forest regeneration is retarded and the condition of the pine roots deteriorates.

Under undisturbed conditions in Lapland the lichen carpet on dry sites of the lichen (Cladonia) forest site type reaches a thickness of 15 - 20 cm. The photograph was taken in Russia close to the mining town of Kovdor and shows a Finnish reindeer-owner, Pekka Välikangas, admiring the thick lichen carpet. Reindeer husbandry is not practised close to Russia's border with Finland.

Colonisation of Finland. Source: Kalliola, 1973.

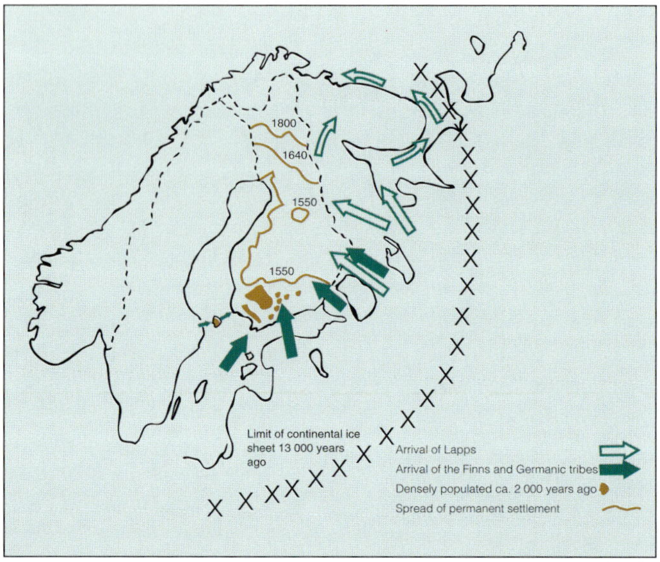

The lichen heaths on the Finnish side of the border have been overpastured. The photograph has been taken in Lapland close to the Rajajooseppi border post on the Russian-Finnish border.

The lichen cover returns slowly after a forest fire. The photograph was taken behind the reindeer fence along the Russian-Finnish border, where reindeer cannot graze. 33 years have passed since the Tuntsa forest fire in eastern Lapland.

The vegetation cover is still almost completely absent in the area grazed by reindeer 33 years after the Tuntsa forest fire in eastern Lapland.

harsh cold climate and forest fire.

The population was small during the hunting, fishing and nomadic period. The availability of food and disease regulated the size of the population. Outside the areas of permanent settlement there were still unlimited expanses of natural forest.

The land grazed by reindeer on this satellite photograph shows up as a dark area on the Finnish side, and a corresponding grey, ungrazed area in Russia. Researcher Kari Mikkola has analysed and produced the picture from a Landsat satellite image.

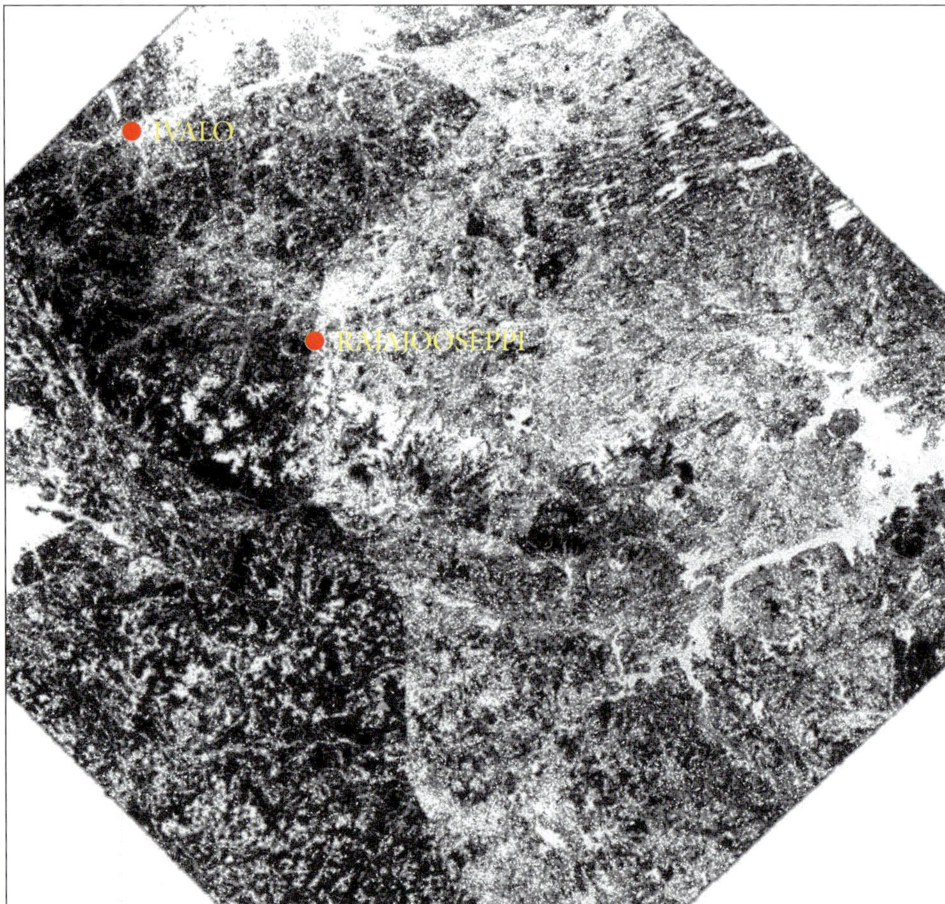

Ungrazed cladonia forest site type (ClT) on the russian side near check point Rajajoo-seppi.

The Lapps have used pine bast as a source of food during famine years. Old trees along the timber line often bear the marks of hard times.

For tar-burning purposes young pines were debarked leaving a strip of bark between the stump and the crown on the northern side. The pines remained alive and large amounts of resin accumulated on the debarked part of the trunks.

Flood meadows along streams and on wetlands were important sources of winter fodder for domesticated animals in the old days. The hay was cut and stored in local barns in July-August, and then carried by sledge in the winter after the watercourses had frozen up. The photograph was taken at Lieksa in northern Karelia.

The signs of old felling work can still be seen on the dry heaths traversed by fire in eastern and northern Finland especially. The large, charred and resiny pine stumps date back to the time of selection felling when only the largest trees in the forest were used. The trees have been felled by axe, probably 150 years ago.

Siberian jay (Perisoreus infaustus) is the traditional friend of the lumberjacks in northern Finland.

THE SETTLERS CULTIVATING THE LAND AND HERDING CATTLE

Some of the people engaged in hunting, fishing and reindeer herding established permanent settlements and village communities. Disagreements and skirmishes between the tribes drove people to the north-east and north in search of new areas to settle in. They brought with them the ability to cultivate the land and herd cattle. Hunting and fishing remained an important additional source of subsistence right up until the onset of industrialisation, especially on the fringes of permanently settled areas.

The battle axe culture appeared on the silt and clay lands in the southern and south-western parts of Finland at the beginning of the Iron Age 4 000 years ago. The slash-and-burn fields and forest pastures supplemented the small permanent fields. The cold period before the start of the Christian era almost completely wiped out the people cultivating the land.

Tacitus the Roman described the people living in Finland in his book "Germania" in A.D. 98:

"Far behind the Germanic tribes were these Finns, who did not own anything and who were completely wild. They had neither horses nor houses, they were dressed in skins, they did not cultivate the land but ate whatever they found, and they slept on the ground. The wretched wicker huts were their only protection against predators and bad weather. They did not have iron, but used bone arrow tips. Equipped in this fashion they hunted in the forests, together with the women, who demanded their share of the spoil."

Permanent settlement and clearing forest for fields became widespread in the Iron Age during the first decades of the Christian era. Inland areas were considered safer than sites along the coasts, which were exposed to the ravages of ship-born raiders and mercenaries.

By the end of the ca. 1 000-year period before the birth of Christ, the area permanently settled by the small population extended to around latitude 62°. The main settlements were in south-west Finland, in the area drained by the River Kokemäenjoki, in southern Häme and in the part of Karelia adjoining Lake Ladoga. Settlement of central Finland started in the middle of the 16th century. Settlers reached northern Finland during the next century. The rivers and lake chains, where the most fertile agricultural land was also to be found, were the main migration routes.

The tribal areas with their central villages formed the backbone of settlement. Isolated homesteads set amongst the forest have always been an essential part of the countryside. Settlement in eastern Finland was centred on the hills; the hilltops were fertile because they had remained above the water after the Ice Age. The isolated holdings were like the antennae of society, forever seeking for fertile pockets in the ever colder and more barren conditions, running from the intense competition for fields and pastures in the populated areas. The villages divided the unpopulated backwoods into hunting and fishing areas.

The bases used on hunting trips often became pioneer farms. During the Swedish-Finnish period the state encouraged, by means of tax-free years, the spread of settlements to the east and north-east towards the region governed by Russia's sphere of influence.

Man became the major factor changing the forest and its ecosystems already during the time of muscle power. The consumption of wood was great and very wasteful. In addition to buildings, almost all the tools, utensils and vehicles were made of wood. Wood was burnt almost continuously in inefficient fires during the winter. Inland settlement was based on slash-and-burn agriculture and it was for long, in addition to tar making, a major consumer of wood that is fully comparable with current levels of industrial wood consumption.

The land suitable for slash-and-burn was often covered with a dense spruce stand. The small trees were felled and the largest trees, about 300/ha, were notched and girdled to kill them. When the trees were dry, the brush and trees were burnt and the ground harrowed with a branch harrow. A lot of ash was formed owing to the high tree density of natural forests. In those days the yield obtainable from burnt forest land was as great or greater than that from the permanent fields. Rye was usually grown first, followed by barley and turnips. A few years after the burn the area was left to be afforested or, if located close to settlements, used as pasture.

The new forest which developed after slash and burn consisted of broad-leaved trees and pine. In rotation cultivation the trees were left to grow for 15 - 30 years, by which time there were sufficient trees to repeat the burn. Repeated slash-and-burn and grazing resulted in broad-leaf dominated forest. Conifers, especially spruce, were removed in order to improve the growth of herbs and grasses.

As the population increased "barley bread" slash-and-burn started to be practised in young broad-leaved and mixed forest. The slash-and-burn rotation shortened to 10 - 15 years. The number of trees available for burning became so small that wood had to be brought in from the surrounding forest. As the time left for the forest to grow shortened, the nutrient reserves of the soil no longer had time to replenish themselves. Slash-and-burn agriculture became over-exploitation and the yields started to decrease during the 18th century. So much forest was destroyed that in 1734 a forest statute was enacted which forbade slash-and-burn for purposes other than the clearing of fields and grazing.

At its greatest, the area of land used in slash-and-burn agriculture was 4 million hectares, the consumption of stemwood being about 10 million m^3 a year. It has been estimated on the basis of the mean slash-and-burn rotation period and the size of the yields, that the production of food for Finland's present population by means of slash-and-burn agriculture would require about 20 million ha. This is about ten times the present area under agriculture.

The forest was used for cattle grazing up until the first half of this century. It has been estimated that in 1900 there were 311 000 horses, 1.4 million cows, 985 000 sheep and 7 500 goats grazing in the forests, on slash-and-burn fields and on natural grazing land.

Because of grazing the fields and grassland had to be fenced. The length of the fences has been estimated at

Growth of the Finnish population.

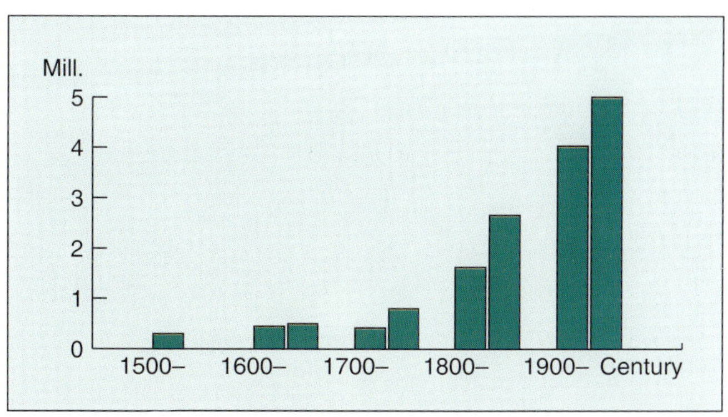

Frequency of slash-and-burn agriculture in Finland during 1700 - 1850. Source: Heikinheimo, 1915.

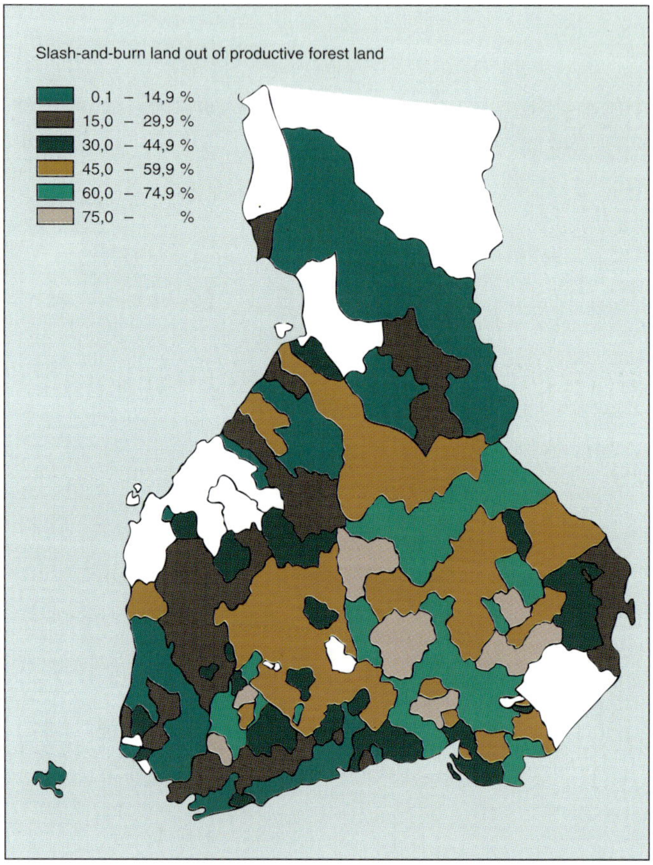

Slash-and-burn was a common practice in old spruce-dominated forest. The biggest trees were girdled in order to kill and dry the trees while they were standing. Old girdling marks can still be seen on old trees, like the one in this photograph taken at Jyrinvaara in northern Karelia.

Forest land ownership, 1982-1992.

Owner	Northern Finland	Southern Finland	Total	of growing stock
	%			
Private	43.6	75.4	61.8	75.0
Companies	5.1	11.3	8.7	9.0
State	46.7	8.0	24.5	12.0
Other	4.6	5.3	5.0	4.0
Total	100.0	100.0	100.0	100.0
Forest Land, 1000 ha	8 575.0	11 499.0	20 074.0	

Area of private forest land owned by	%
Farmers	42
Wage earners	24
Entrepreneurs	5
Pensioners	29
Total	100

The group "othert" includes municipalities, parishes and jointly-owned private forest

800 000 km. To maintain these fences 0.7 million m³ of wooden poles were needed each year. In addition to herbs, grasses and sedges, tree foliage was an important source of forage for cattle. Large amounts of branches and coniferous foliage were used as winter bedding in the cattle sheds and stables. Together with manure and peat, they were used to maintain the fertility of the fields.

During the worst famine years, hunger and disease killed off up to one third of the population. During one decade in the 19th century there were two famine years and three poor years in southern Finland. The most recent famine occurred in 1867-68. In times of famine bread had to be

The majestic hill forests and expansive lake views of the Koli area in northern Karelia make it the national landscape of Finland. Moves are being made to preserve this traditional landscape in the Koli National Park by means of slash and burn. A slash-and-burn field was burnt in the national park in June 1994. The old spruce forest was cut the previous year to dry it out ready for burning.

After slash and burn the land was used for grazing cattle and eventually became covered with alder. There are areas in Kaavi in northern Savo where the signs of slash and burn are still clearly visible in the pattern of land use.

The cessation of grazing promotes the spread of alder. Grey alder has little commercial value and the alder stands growing on fertile sites are classified as unproductive land; the state provides funds for regenerating such land. Spruce was planted under the grey alder stand at Kaavi 10 years ago.

made from natural sources growing in the forest. Pine bast was the most common emergency source of bread. Bast bread was prepared from the bast of pines, felled in the spring, by mixing it with threshing chaff and straw. Other ingredients were wild berries, sorrel, the root pith of rushes, the roots of sedges and bog arum, and linseed.

Spruce was planted in this alder stand during the 1940's. The stone ruins are signs of the earlier use of the land for fields and grazing. The last grey alders died out in the deep gloom of the spruce stand on this Oxalis-Myrtillus type site at Kaavi in northern Savo. The annual increment of the planted spruce stand is estimated to be at least 10 m^3/ha.

This dead and dying slash-and-burn birch stand is infested with rot fungi. The photograph shows a coral fungus (Hericium coralloides) growing on a dead birch. The species is rather rare in Finland.

Wood as a raw material and export commodity in the pre-industrial age

The mining of ore and metal smelting were the first heralds of industrialisation. Wood and charcoal were burnt in mining, in the production of metals and in the manufacture of metal tools and weapons. Wood was used as a source of heat in glass manufacturing, and the ash of birch and especially aspen was used to make potash.

This slash-and-burn birch stand in northern Karelia has regenerated naturally 90 years ago. The site is of the Oxalis-Myrtillus forest type. If succession is allowed to proceed undisturbed, the birch stand will continue to degenerate and spruce gradually take over the area.

Prescribed burning, which destroyed the vegetation and burnt off the humus, has revealed the original use of this piece of land. A family lived in a cottage at the site of these stone ruins during the 1860's. The father moved to America to seek a better life for his family, who were to join him later on. Famine and hunger tragically altered their plans in 1867. A passer-by found two children crying on their dead mother's breast, and took the children to safety. Over the years the building rotted and caved in, and the forest gradually took over. The present forest owner at Töysä in southern Ostrobothnia has left the ruins to stand as a monument to the tragedy.

Tar was the most important product of the forest in pre-industrial times. During the age of wooden ships it was in great demand both at home and abroad. The rural population obtained urgently needed income by burning and selling tar. During years of famine and when game was scarce the income from tar was essential for survival. The buyers and exporters of tar built flourishing towns and ports at the mouths of rivers along the coast of the Gulf of Bothnia.

Outside the immediate vicinity of populated areas tar burning, together with slash-and-burn agriculture, had the greatest effect on the forests. The young and middle-aged pine stands growing on dry, warm mineral soil provided the best raw materials for tar burning. The river and lake networks formed the main transportation routes. The major tarproducing areas were in Ostrobothnia and Kainuu, and along the rivers in southern Finland. Up until the 1830's tar was the most important export commodity.

It has been calculated on the basis of the documented number of tar barrels that, when tar production reached its peak during the 17th and 18th centuries, at least 10 million m^3 of pine were utilised annually. The pulp industry consumed as much during the 1970's. The wastewood left in the forest and the fires that broke out from the tar pits increased the forest drain to a level considerably above that which was consumed in production proper.

The first water-driven saws were established in 1533, and became common during the 1600's. However, it took a long time before the export of sawn timber exceeded the exports of spars, boards and deal cut by axe and handsaw. Building timber, pit props and firewood were also important export commodities.

Sawing was considered to be a threat to the forest resources in the first half of the 19th century. The efficient steam saw was especially feared. A permit for establishing the first steam saw was granted in 1857. Liberal economic policy and the growth in the demand for sawn wood in Europe resulted in an increase in the number of sawmills. Steam produced by burning wood foreshadowed the onset of efficient industrial production. It also revolutionised transportation. The first machine shop was established in 1837, its first product being a steam engine. However, a regular shipping route had already started on Lake Saimaa using a steam engine purchased from abroad. Steam was used to transport timber and wood products from inland locations to the cities and ports on the coast.

Industrialisation in the true meaning of the word started when the burning of wood for energy was gradually replaced by imported fuels such as coal and oil, and electricity produced by water power. Wood became the raw material of a diversified forest industry.

The spread of literacy and the development of printing techniques considerably increased the demand for paper. A shortage of the cotton and rags used as the raw material for paper, as well as their high price, prevented an increase in consumption. Paper was imported to Finland in the first half of the 19th century.

The situation changed when paper started to be made from wood fibre. The forest resources attained the value which they still hold today. The first mechanical pulpwood mill was started up in 1860, the first sulphate pulp mill in 1880, and the first sulphite pulp mill in 1885. Finland discovered its green gold and industrialisation began.

The use of low-value tree species such as alder (Alnus spp.) as fuelwood on farms also promotes the silvicultural treatment of the forests. Fuelwood consumption at the beginning of the 1990's was around 3 million m³, compared to 15 million m³ 30 years earlier.

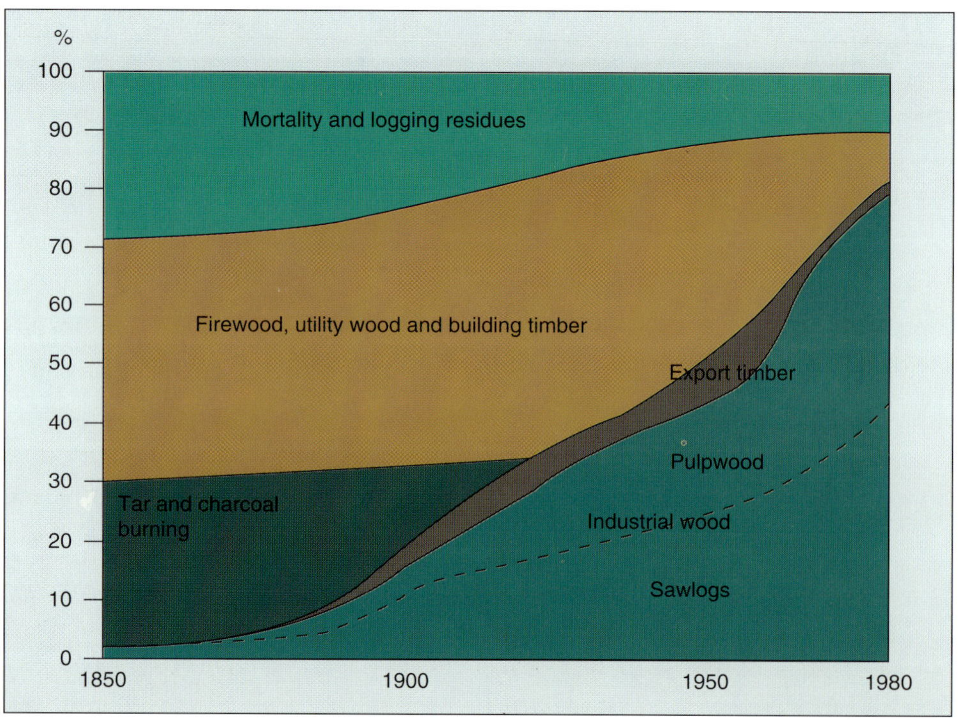

Relative distribution of the stemwood removal according to use and mortality and logging residues during 1850-1980. The values are approximations up to the 1920's.

The forest as a source of wealth for social development

The growth of the forest industry into a major export sector accelerated the development of agriculture and dairy farming, industrialisation, diversification of production and material prosperity, and formed a foundation for development of the population's standard of living, culture and social security.

Industrial utilisation gave wood an unprecedented value, and timber harvesting and transportation provided work in all parts of the country. The income from wood sales and forest work has been of great importance for the rural population because farmers in Finland have always been, as is the case in the other Nordic countries, independent owners of land, forest and production equipment.

As industrialisation progressed the position of the agricultural population was strengthened by the granting of holdings to the landless rural population. Forest owned by the state and forest companies was primarily appropriated for this purpose. The central units of the forest ownership structure are the families who nowadays obtain 80-85 % of the total stumpage income from the sale of timber.

The increase in the proportion of industrial wood out of total consumption increased the financial return. Of the total removal of stemwood halfway through the 19th century, about 3 % was industrial wood, 27 % was used for slash-and-burn agriculture and tar and charcoal burning, 40 % for firewood, utility wood and construction timber, and 30 % was left to rot in the forest as cutting residues or mortality. In 1920 the proportion of industrial wood was 20 %, compared to 82 % nowadays. At the present time the total removal includes 8 % firewood and utility wood, and 10 % logging residues and mortality. Changes in the structure of wood utilisation have been speeded up by the replacement of firewood by other energy raw materials and by electricity.

The importance of forestry work was emphasised by its concentration in that part of the year when it was not possible to work in the fields. The timber was harvested during the winter and transported by horse and sledge to the banks of the rivers and lakes. The wood was floated on the spring meltwater, the floating work stopping when it was possible to get on with the field work.

While the forests and industry provided work and income for agricultural investments, the location of farms amongst the forests has meant a considerable advantage for the Finnish forest industry compared to those parts of the boreal coniferous zone where most of the forests are state owned and lie outside the settled areas.

The farms provided labour and horses for forest work up until the middle of this century. The rural road network facilitated the supplying of forest work sites and timber transport. The removals in farm-dominated forests has been 2-3 times greater than that over extensive areas of the boreal coniferous forest zone. Silviculture has maintained a high level of tree growth.

The forest industry was located in the mouths of rivers along the coast, and inland close to sources of water power. The factory places provided work, and subsequently developed into growth centres with a diverse manufacturing base.

The forest industry has procured exports income for the national economy. In the early stages of industrialisation exports were needed to finance the import of raw materials, machinery and equipment as well as foodstuffs.

Forest exports have for long been irreplaceable owing to the scarcity of other natural resources.

Exports of roundwood and hewn timber were considerable in the early days of the forest industry, as well as later on during the period of reconstruction after the world wars. Sawn wood, plywood, wood products and pulp accounted for 79 % of the total exports of the forest industry at the end of the 1930's.

The proportion of paper and paperboard out of the value of exports increased from 30 % to 68 % during the period 1955-1992. The production and export of writing paper and newsprint especially increased rapidly. The degree of processing of Finnish forest products is higher than that in any of the other major exporting countries.

The decisive importance of exports for the economic development of independent Finland is demonstrated by the fact that the total real value of exports increased 19-fold from the 1920's to the 1990's. The forest industry's share of the value of exports was about 80% up until the 1950's. Following the diversification of production it has fallen to about 40 %. Despite the decrease in the relative share, the real value of the forest industry's exports increased 10-fold from the 1920's up to the 1990's. This rising trend tailed off during the 1980's when the supply of wood and the price competitiveness of the industry deteriorated. A new growth period started in 1993.

The metal-working, engineering, shipbuilding and electronics industries have grown as a part of the forest industry and in the allied industrial sectors. In relation to the size of its population, Finland is a major manufacturer of machinery for the forest industry and forestry, as well as an important exporter of machinery and know how in this field. The forest industry and associated metal-working and engineering industries and teleinformatics form Finland's most competitive manufacturing and know-how cluster.

The great inland lake basins are ideally suited for bundle floating, which accounted for 10 % of the long-distance transport of timber in 1993. Road transport was the major form of transport (62 %), followed by rail transport (28 %). Loose floating stopped on the major rivers in northern Finland at the beginning of the 1990's. The photograph was taken in the harbour of Kymmene Oy's pulpmill in the southern part of Lake Saimaa.

Exports of the forest industry played a highly important role in rebuilding the national economy after the world wars. During boom periods it has been the driving power behind the national economy. A down-turn in the growth of this sector has preceded national slumps. Owing to its high share of domestic input, the forest industry still procures at least 40 % of the country's net export earnings. In its role as a wood purchaser the industry provides work and income and helps keep the rural areas populated at a time when the overproduction of foodstuffs is causing a cutback in arable and cattle farming.

Measuring sawlogs and checking their quality are nowadays highly automated procedures. The logs arriving at the sawmill are measured and graded for production purposes, as well as to determine the transport costs and the price to be paid for the timber. The photograph shows a spruce sawmill at Renko in southern Finland.

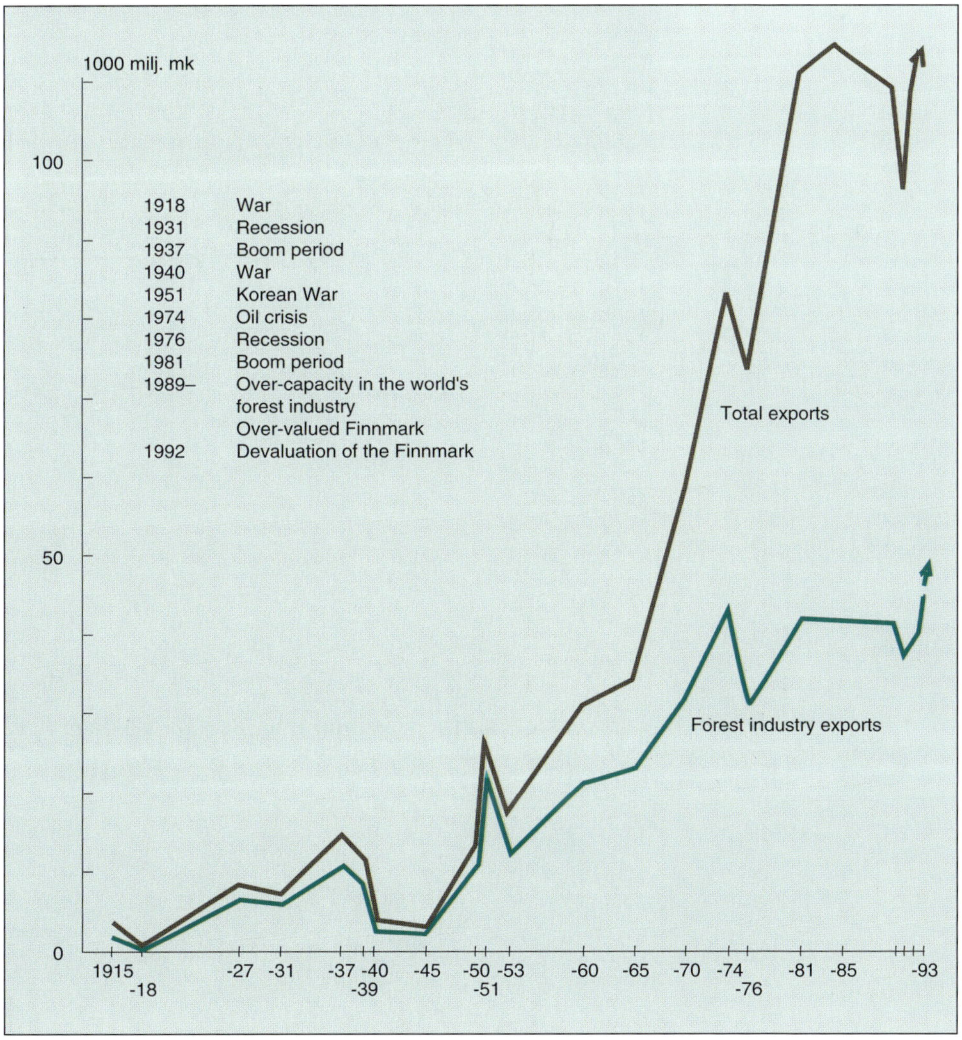

Total exports and the real value of forest industry exports in 1992 monetary value during 1915 - 1993.

Structural development of the utilisation of removals

Type of use	1947–49	1979–81	1989–91
	%		
Industrial wood	41.2	86.9	90.2
Roundwood export	6.4	3.8	1.5
Fuelwood	39.0	7.9	7.0
Other	13.4	1.4	1.3
Total	100.0	100.0	100.0
Total million m^3/y	41.0	51.8	46.8

Development of the utilisation of roundwood and pulpwood by the forest industry

Type of roundwood	1947–49	1979–81	1989–91
	%		
Domestic roundwood	96.0	78.0	76.8
Imported wood	–	6.9	10.7
Wood residues	3.4	14.0	10.6
Recycled fibre	0.6	1.1	1.9
Total	100.0	100.0	100.0
Total million m^3/v	17.6	57.7	57.3

Development of the production of forest industries and the export shares by products

Product	1938	1950	1980	1993
Production				
Sawn goods mill.m^3	6.0	4.2	9.4	8.3
Wood panels mill. m^3	0.3	0.3	1.6	1.2
Paperpulp mill.t	2.2	1.6	7.2	9.3
Paper mill.t	0.6	0.6	4.5	7.8
Paper-board mill.t	0.1	0.1	1.5	2.2
	1938	1955	1980	1993
Proportion of the value %				
Sawn goods	34.6	28.7	22.1	13.3
Wood panels and other	8.4	11.7	9.6	8.0
Paperpulp	35.9	28.9	15.4	6.4
Newspaper	11.4	11.1	10.8	6.4
Kraft paper	0.9	4.1	3.2	2.8
Printing and writing p.	} 5.3	} 8.7	17.3	39.8
Other paper			2.9	3.5
Paper-board	3.4	5.6	10.6	14.6
Processed p. products	0.1	1.2	8.1	5.2
Total %	100.0	100.0	100.0	100.0

Forestry components and their share out of national product in 1990.

Net stumpage income	6 612	Mmk
Income from timber procurement	589	"
Contractors' income from logging	595	"
Forestry promotion	220	"
Deterioration of capital	2 164	"
Wages and salaries	2 339	"
Other items	539	"
Total	13 058	"
Proportion out of total national product	2.9	%

Forest cluster developed around wood processing industry. Its exports account for about 70 % of the value of total exports.

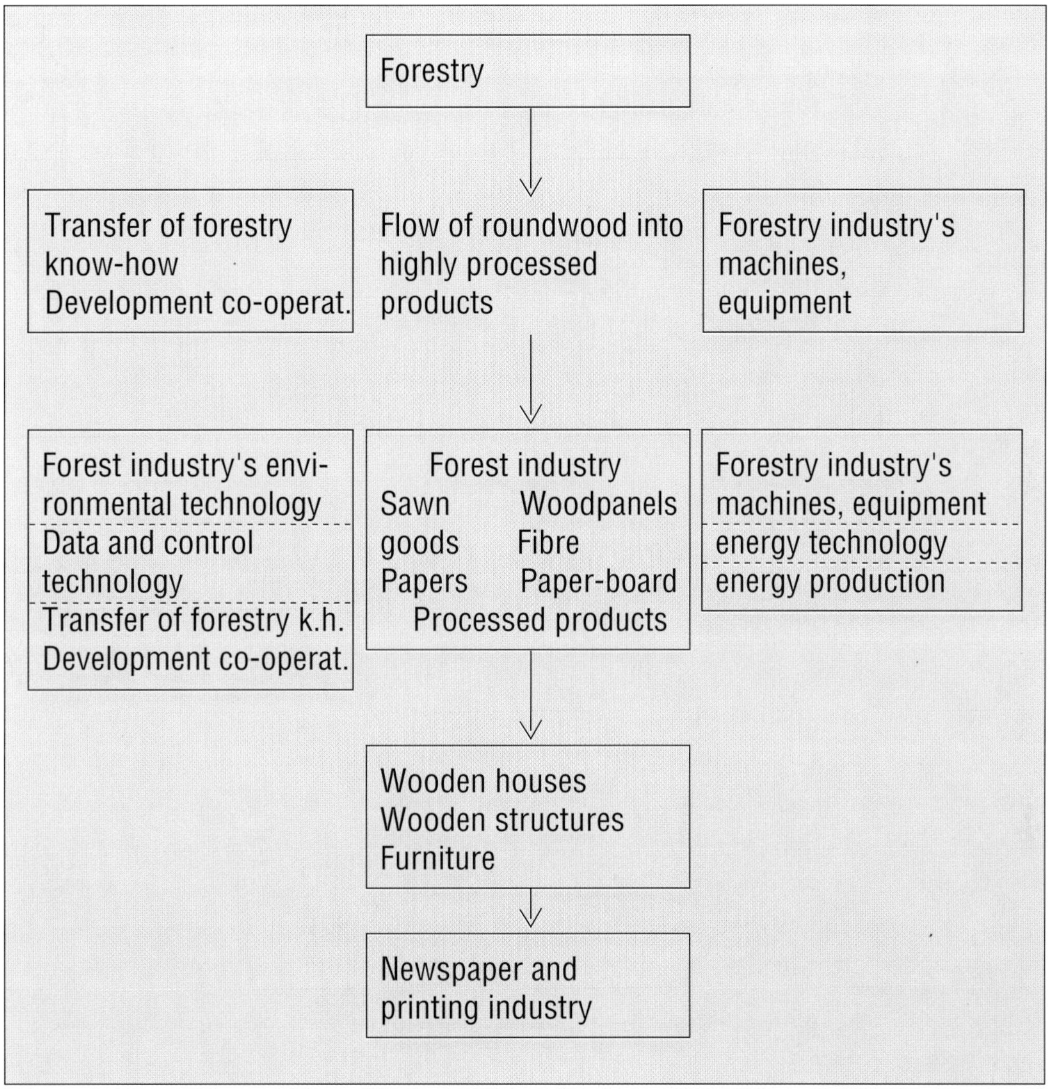

	1965	1970	1980	1990
Million mk	6 284	7 140	9 112	7 303
%	12	16	15	19

Stumpage incomes in 1992 monetary value and the proportion of the costs of silviculture, basic improvements and road construction out of stumpage incomes.

Permanently settled areas require services and a road network, which also benefit forestry in northern and eastern Finland. Forest income is essential for the inhabitants of these regions, and forestry is dependent on the presence of a local supply of labour. The photograph was taken at Lieksa in northern Karelia.

THE FOREST AS AN ENERGY RESERVE

Wood was almost the only source of energy up until the 19th century when water power started to become available. The use of coal and oil became widespread at the turn of the century. Wood still supplied about 80 % of the basic energy requirements at the beginning of the 1930's. During the Second World War its share increased to 85 %. The annual consumption of firewood peaked at 18 million m^3, and exceeded the consumption of wood for industrial purposes at the end of the 1940's.

The production of basic energy increased from 7 million equivalent oil tonnes in 1950 to 30 million tonnes in 1990. The shares of coal and water power initially increased. The share of oil reached a peak in the 1970's. The combustion of wood billets decreased sharply and was 3.3 million m^3 at the end of the 1980's. At the present time, however, it is increasing. The oil crisis in the 1970's initiated a search for alternative raw materials for energy production, which was partly met by an increase in the use of natural gas and nuclear power, and imports of electricity.

The use of bioenergy, which is primarily based on wood combustion, has maintained its important position. In the 1950's the forest industries made considerable efforts to increase the efficiency of wood utilisation and to decrease the effluent load on the watercourses. Almost all the bark and other wood waste, as well as the residues from cellulose and pulp cooking, are nowadays used for energy production. Around 4 million equivalent oil tonnes, equivalent to 21 million m^3 of stemwood, are annually obtained from waste products and the indirect combustion of wood.

The structural change in the forest industry's use of wood has made a considerable contribution to attaining sustainable development. The proportion of net carbon dioxide emissions, derived from coal and oil combustion, fell from 43 % to 7 % of the total consumption of energy raw materials during 1973 to 1992. Wood residues and nuclear and hydroelectric power are nowadays the most important sources of energy.

Rapid strides have been made in developing bioenergy production technology and in its use. A modern pulpmill is also a bioenergy production plant which sells one third of the energy it produces to other consumers. Power plants burning bark and wood are either integrated with sawmills or located separately. Combustion techniques also permit the use of all the biomass components, as well as peat and other types of raw material. The emission of exhaust gases and other pollutants has decisively decreased.

The forest resources, and peat and hydroelectric power, can meet the country's energy requirements in times of crisis when import routes are severed. The annual stem volume increment corresponds to about 15 million equivalent oil tonnes, and the biomass increment of the whole forest to 23 million equivalent tonnes. However, the present and future prices and costs of the different raw materials make it economically more viable to produce export commodities from wood than to use it as an energy source. On the other hand, under current conditions where the industrial consumption of wood is smaller than the sustainable cutting potential, it is recommendable to increase the combustion of small trees harvested in silvicultural thinnings, as well as wood of less value, as an industrial raw material.

Distribution of basic energy production during 1930-1990.

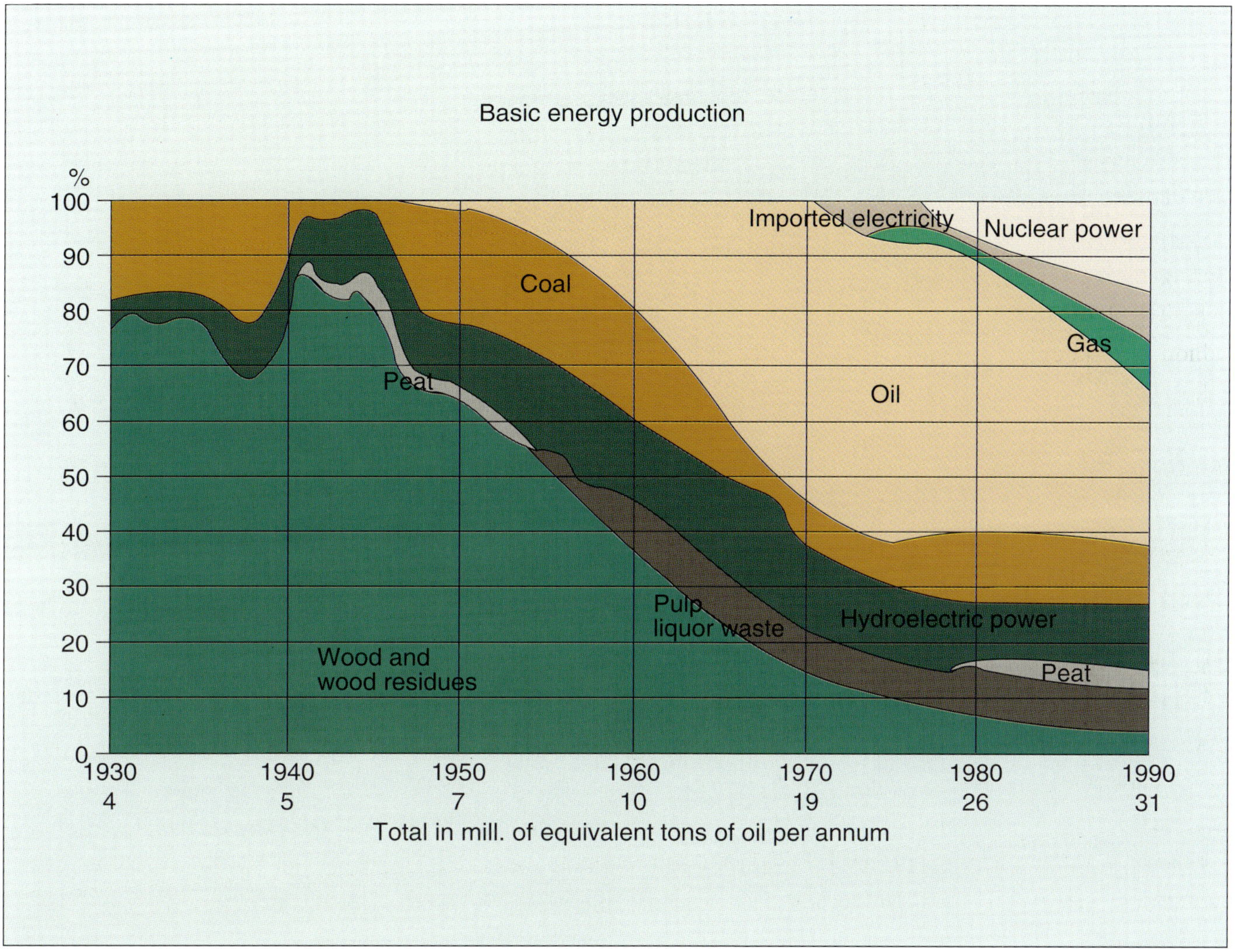

	1973 %	1992 %
Oil	37	3
Coal	6	4
Gas	–	10
Bark and other wood residues	37	37
Peat	–	3
Hydropower	8	16
Nuclier power	–	27
Other purchased electricity	12	–

Development of the energy structure of the forest industry.

Cultural forest

Fields, pastures and parks form the cultural landscape. Their beauty is the balanced result of a landscape created by the long-term impact of man. In areas where the original forest has been almost completely superseded by other forms of land use, virgin forest is considered to be wild and unorganised nature and the antithesis to cultural landscape. In countries where silviculture has long traditions, regeneration areas, seedling stands, dense middle-aged forests and stands of fine sawtimber ripe for regeneration, are held in as high esteem as the other components of the landscape created by man.

The impact of Finnish silviculture in moulding the landscape is of short duration compared to the long growing period of trees. Forest-destroying cuttings and the threat of over-exploitation of the forest resources have been eliminated during the latter half of this century. It has been possible to create balanced cultural forest in parts of southernmost Finland. Planted and sown forest accounts for about one quarter of the area of forest land. For this reason many of the signs of forest management appear to be "the destruction of nature and the antithesis of culture" to a viewer not capable of recognising the aims of long-term forest management, nor the dynamics of the forest.

The contradiction inherent in landscape appreciation is well illustrated by the fact that the forest depicted in classical landscape art at the end of the 19th century and beginning of the 20th century is called cultural landscape. However, it consisted primarily of slash-and-burn openings and coppiced forest. Its appreciation was based on an awareness that the landscape was the result of the exploitation of natural resources essential to the livelihood of man.

This should leading to a balance of the appreciations currently under public debate, in which the natural romanticism of post-industrial man is set against the silvicultural measures essential for his prosperity. When the beauty of the countryside is seen as an equilibrium between man-made fields, pastures and forest, i.e. the end product of man's work and the forces of nature, then cultural forests will be accepted as a value and an objective.

A drained dwarf-shrub pine mire changes into a cultural biotope where the growth of pine is as good as that on corresponding mineral soil sites. This drained pine mire, which has changed into dwarf-shrub heathy peatland, is located in one of the oldest drained areas in southern Finland. The stand volume was about 50 m^3/ha when the Jaakkoinsuo mire at Vilppula in northern Häme was drained for the first time in 1909. The thickness of the peat layer is 2.5 m and the stand 200 years old. The stand volume is about 180 m^3/ha and the increment 1.8 m^3/ha/year.

Pine mires are peatlands of medium or low fertility with a tree cover of pine and/or pubescent birch. The natural dwarf shrub pine mire shown in the photograph is a peatland type that is still considered worth draining in southern Finland. Pine mires and their drained counterparts account for almost 60% of the peatlands in Finland.

FROM THE UTILISATION OF NATURAL FORESTS TO SUSTAINABLE AND PROGRESSIVE FORESTRY

THE UTILISATION OF NATURAL FORESTS

In the old days, wood was harvested whenever necessary from natural forests that were located as close to the site of consumption as possible. When certain types of wood, e.g. trunks for building log cabins, became exhausted in the vicinity of the settlements, the distance to the cutting sites gradually increased. However, the physical effort involved and the location of water transportation routes set limits on this. As the availability of suitable stands diminished and the transportation distances increased, a compromise had to be made with respect to tree size and quality. All sizes of stem could be used for firewood, although birch was preferred over other tree species owing to its high thermal content.

The idea of promoting forest regeneration through tree growing and timber harvesting was not known at the time. This was not necessary in the days when the population was small and the forest resources infinite. Although the area of forest decreased, growing trees remained an alien concept to the farmers whose main use for the forest was a site for clearing fields. However, they knew that any field or pasture left untended for a short period even had to be protected from reforestation. Field clearance was supported with state funds right up until the 1980's.

As the industrial use of wood increased, trees which met the size and quality criteria for sawlogs were harvested. Small, poor-quality trees, as well as large amounts of cutting residues, were left in the forest in areas where they could not be used as firewood or utility wood. The best quality trees in the stand were also harvested for spars, pit props and, to an ever increasing extent, for pulpwood.

For a long time the cutting of industrial wood was carried out as selection fellings according to a minimum diameter limit; all those trees which met the size and quality norms were removed from the forest. The remaining trees were left to grow. As the demand for wood increased, the size and quality norms were lowered.

An attempt was made to further develop the selection felling method by increasing the time interval between the harvesting of sawtimber trees. This would have given even the smallest trees time to grow into sawtimber trees. The forest was supposed to regenerate itself in the openings left by cuttings. However, this was not successful owing to the characteristics of the tree species in the boreal coniferous forest zone. The result was creamed forest which, on land suitable for pine, gradually developed into slow-growing spruce stands. Creamed stands later had to be regenerated by sowing or planting.

Harvesting that was indifferent to the continuity of tree production did not as such destroy the forest. The seed and root suckers of pioneer species, as well as the spruce growing as an understorey, regenerated the forests. In addition to the decrease in their technical quality, the genetic quality of the trees also deteriorated.

Cuttings, slash-and-burn agriculture and tar burning altered the tree species composition of the forests. Conifers decreased and broad-leaved trees, especially birch and alder, increased on the more fertile soils. Spruce occupied the infertile sites normally covered with pine.

APPROACHING EXHAUSTION OF THE FOREST RESOURCES

The drastic reduction in the volume and quality of the forest resources resulted in a shortage of all timber assortments, as well as firewood, in extensive parts of southeast Finland and along the Ostrobothnian coast in the

Estimate of the sufficiency of the forest resources in 1850.
Source: Laitakari, 1960.

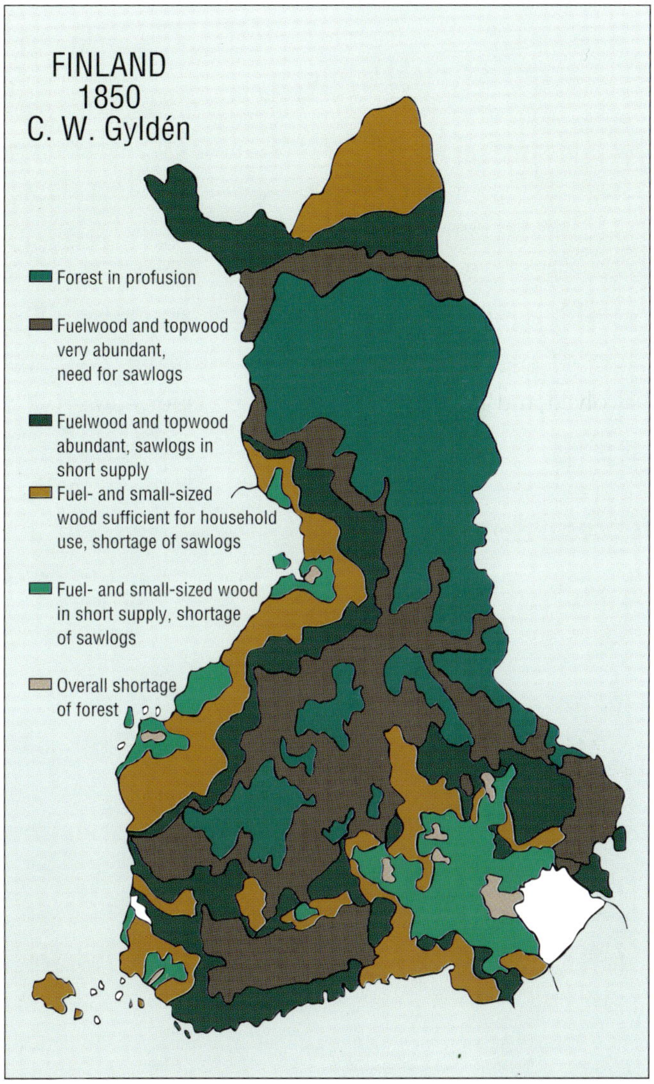

middle of the 19th century.

Foreign experts were invited to Finland to evaluate our forests and to make proposals about how forestry should be developed. The situation is well illustrated by the report of Prof. Edmund von Berg, head of the Tharandt Forestry College in Germany, for the Finnish senate in 1859. He wrote:

"The destruction of the forests, which the Finns have become rather adept at, is further promoted by the free grazing of cattle, slash-and-burn agriculture and very damaging wild fires, or to put it in more precise words: these three means were all being used to achieve the same noble aim, namely the destruction of the forests."

"The untended, destroyed or burnt forests to be found in Finland have made me very sad and rather downhearted. In fact although I did not approach the Finnish forests with very high hopes, I did not expect to see such great destruction. Only the stupidest person could regard this with tepidity."

Estimates of the forest resources and the level of timber consumption were based on observations made by officials from the National Board of Survey and the National Board of Forestry, as well as on surveys of the volume and quality of the growing stock in restricted areas. The first forest balance in which estimates of the volume and increment of the growing stock and the wood drain were compared was completed in 1873. According to this, the annual under-bark increment in stemwood volume was 31.5 million m^3, and the wood drain 29.5 million m^3. In the forest balance completed in 1913 the increment estimate was 35 million m^3 and the drain estimate 40 million m^3. The drain included 12.5 million m^3 of industrial wood.

The results of the latter forest balance were taken to indicate overcutting. The situation was made worse by the fact that, in those days, much of the natural forest was inaccessible to cutting and transportation. Although the results of the first forest inventory made using statistical sampling methods, which came out at the beginning of the 1920's, indicated that the growth of the standing crop had been underestimated in the first forest balances, the ever-increasing consumption of wood by the forest industry presupposed an improvement in cutting methods and in silviculture.

Development of forest legislation and administration

The reduction in timber resources located within the areas where harvesting and transportation were possible resulted in legislation and the development of administration aimed at ensuring the supply of wood to the mining and metal-working industries and shipbuilding. The cutting of hardwoods for home use and slash-and-burn agriculture was prohibited.

Fears were expressed at the beginning of the 19th century when water-driven saws were becoming more common, and the consumption of wood was high and wasteful in slash – and – burn agriculture, tar burning and household use, that the forest resources would be exhausted. The time of forest committees, legislation and the development of administration had arrived.

The Central Administrative Board of Land Survey and Forest Management was established in 1851. The Finnish Forest and Park Service became a separate agency in 1864. The main task of the reformed administration was to continue the development of the land ownership structure, started in the 18th century, and to designate the "excess land" in public use as state land.

The forest administration initiated the spread of silvicultural practices, the procurement of wood for industry, and gradually stopped the harvesting of free wood by the local population in the state forests. At the beginning of the 1920's the Finnish Forest and Park Service had its own sawmills and pulpmills in northern Finland. These subsequently developed into the major industrial enterprise nowadays owned by the state in Kemi. Up until the 1980's the tasks of the Finnish Forest and Park Service also included supervision of the forest legislation, non-academic forestry education and log-floating. Nowadays the institution is a state enterprise engaged in timber production in the state-owned forests. Other major tasks are the management of a number of nature conservation areas and national parks, and the promotion of multiple forest use on the land and watercourses under its jurisdiction.

The Forest Act of 1886 obliged the forest owner to ensure natural regeneration of the forest after final cutting. In 1927 this was developed into the Private Forest Act which, with slight modifications, is still in force. It applies to all the forested areas in Finland, apart from state-owned forest, and prohibits the destruction of forest.

Forest destruction refers to the sort of final cutting which endangers the development of a naturally regenerated seedling stand capable of development, and the sort of thinnings which do not conform to the silvicultural principles applied in growing tree stands. Final cutting can be carried out in stands whose age and diameter are equal to or exceed the norms laid down in the legislative guidelines.

If a forest owner violates the law, the cutting work is interrupted by the authorities, who then try to come to agreement with the forest owner about the restoration or regeneration of the destroyed part of the forest. If agreement is not reached, the forest is placed under a protection order for a specific period of time, usually for 10 years. During the protection period the forest can be used in accordance with a plan approved by the authorities. If the owner does not restore or regenerate the destroyed forest, the necessary work is carried out by the authorities and the owner is charged for the work.

The act only covers commercial fellings. There is no obligation about managing forests where commercial cuttings are not performed, nor about selling stands ready for final cutting. The act ensures the sustainability of timber production.

The Finnish Forestry Society was founded in 1877. It has since developed into the main ideological and professional organisation in the field of forestry, its members including professional organisations and the institutes of the forestry administration and interest groups. The tasks of the society include the promotion of forestry, co-operation at home and abroad, especially within the Nordic countries, and the dissemination of information.

In 1950 an act was passed about forestry-promoting associations in the private forestry sector. Membership of the associations is voluntary, but every forest owner whose mean stemwood yield, estimated for taxation purposes, is at least 20 m^3 must pay, according to the act, a silvicultural levy. Elected representatives make decisions concerning the management of the associations and also about the silvicultural levies. The size of the levies is defined in the act, and they have ranged from 2 to 6 % of the taxable yield of the forest. The other income of the associations comprises fees for forestry work and services.

Since the 1920's the taxation of forest income and forest capital has been based on the mean stemwood increment on forest land of different yield classes, on the timber assortment structure of the cutting potential, and on mean stumpage prices. The system has encouraged good forest management and timber sales. Since 1993 the forest owners have been able to choose either taxation based on timber sales or to continue with the earlier areal taxation system for a 13-year transition period, after which all taxation will be based on the income from timber sales.

The forestry board organisation was established in 1917 to supervise the Private Forest Act and to promote private forestry. Nowadays it comprises 19 forestry board districts and the forest department of the self-governing province of the Åland Islands. The activities of the boards are directed by two forestry centres: Forestry Centre Tapio in the region comprising the 17 Finnish-speaking forestry districts, and Forestry Centre Skogskultur in the region covering the two Swedish-speaking forestry districts.

Representatives elected by the forest owners have a majority on the executive board of the forestry centres and district forestry boards. In addition to the representatives of private forestry, the state, the forest industry and forest workers also have their own representatives. This self-governing organisation, whose tasks also include supervision of the Private Forest Act, has proved to be highly effective. The state finances most of the running costs of the forestry board districts and the forestry centres.

The Ministry of Agriculture and Forestry is also responsible for international co-operation. The administration of the forestry board districts and related forest legislation are currently being renewed in order to correspond better to the needs of modern society; in addition to timber production, the aim is to maintain multiple use values and the biodiversity of the forests.

Private forest holdings, number of them 436 982
forest land, 11.5 mill. ha
Paying silvicultural levies, number of holdings 290 814
forest land 11.0 mill. ha

Forest holdings paying full silvicultural levies
Area of forest land in the forest holdings, ha

<5	5-19.9	20-49.9	50-99.9	100-199	200-500	>500	Total
Proportion, % of the number							
2.2	42.8	34.4	14.6	4.9	0.9	0.2	100.0

Number and area of private forest holdings and the distribution of the number of them paying full silvicultural levies into size classes.

Research and education

Political economists were interested in forestry already during the 18th and 19th centuries. Officials from the Administrative Board of Land Survey and Forest Management made estimates of the forest resources and their sufficiency. Timber harvesting methods which promoted natural regeneration of the forest were studied.

Wood production methods based on the growing of forest were adopted in the forests of the state management areas and official residences, as well as on the large privately owned forest estates. They formed the topics of the fledgling research and education activities, and provided a model for other forest owners.

The education of forest officers was started at Evo in the state management area located about 120 km to the north of Helsinki in the 1860's. The teachers at the school investigated silviculture and tree harvesting. The head of the school, A.G. Blomquist, published the first growth and yield tables for pine, spruce and birch in 1872. The first silvicultural handbook appeared in 1853.

Forestry education was transferred to the University of Helsinki as part of the Faculty of Agriculture and Forestry in 1908. The Finnish Forest Research Institute was established in 1917. The institute was given state forest for research purposes in different parts of the country. Organised forestry research and professional and scientific forestry education started in Finland at a time when the country was gaining independence. The institutes and departments in a range of scientific disciplines, chaired by professors, formed the backbone of education and research. Silviculture, forest soil science, forest mensuration and forestry planning, forest biology, peatland forestry, forest technology and forest economics were the first main fields.

Education and research expanded rapidly after the 1940's along with the increase in forest utilisation and the development of work technology and data processing. The number of professorships, researchers and lecturers increased. New disciplines such as forest genetics, forest pathology, forest zoology, environmental science, timber harvesting, wood technology, business sciences, bioenergy research, forest products marketing, land use economics and private forestry now have their own chairs.

Owing to the great importance of forestry products for the domestic and export markets, a wood marketing department was established in the Faculty of Agriculture and Forestry at the University of Helsinki in 1944, and a professorship in this field in 1959. The international standing of wood products marketing and other fields has created a basis for the unification of Finnish forestry with economically integrating Europe.

Development in the field of forestry research is well illustrated by the tenfold increase since the 1940's in the number of researchers at the Finnish Forest Research Institute to the present strength of about 220. Expansion of the institute occurred partly through the establishment of a number of regional research stations during the 1960's. There are now 8 stations, with more than one third of the institute's researchers working at them. The institute has 140 000 ha of research forest, of which about 80 000 ha are used for commercial timber production and 60 000 ha for nature conservation areas and national parks.

A faculty of forestry was established at the University of Joensuu in 1982. The European Forestry Institute, which is also located in Joensuu, was founded by Finland as an independent international research body in 1992. Its task is to carry out problem-oriented multi-disciplinary research, the results of which serve forest policy decision-making in Europe.

In addition to the universities, there were 25 colleges and schools turning out forestry engineers and forestry technicians in 1993. Training is also provided to forest workers, forest machine users and forest owners. The Forest Work Study Section of the Association of Finnish Forest Industries is responsible for developing timber harvesting and silvicultural techniques. The Work Efficiency Association develops working methods for farm forestry. Process technology in the forest industry has its own research institutes and laboratories.

In the early days of forestry education many researchers and teachers learnt about silviculture in Central Europe, especially in Germany. However, their methods could not as such be applied directly to Finnish conditions. It was considered that sustainable results could only be obtained by taking into account the climate and terrain of the country, the structure of forest ownership and the development stage of Finnish society.

The individuality of Finnish silviculture is well illustrated by A.K. Cajander's forest site type theory. Cajander was a botanist, the head of the Evo Forest School, the first professor in silviculture at the University of Helsinki, the director general of the Finnish Forest and Park Service, and eventually the country's Prime Minister. He planned and organised forestry education, research and administration from the 1910's to the 1930's.

According to the forest site type theory, the forest and peatland types defined on the basis of the ground vegetation, reflect the fertility of the soil and wood production capacity, i.e. productivity, of the site. This method had already been employed by farmers when looking for land suitable for slash-and-burn agriculture, pastures and fields.

Every ground vegetation species has its own specific site requirements. The plants compete amongst themselves for light, nutrients and water. The species either support or try to suppress each other. Plant communities develop towards a state where the species composition is the same on sites of the same productivity, and different on sites of different site productivity. If a disturbance in the ecosystem results in the disappearance of certain species, they return to their own forest site type.

The forest site types are defined on the basis of the ground vegetation, especially indicator species, growing under a fully stocked, mature tree stand. The use of forest site types presupposes that the plant species and species composition in all stages of forest succession are known.

The forest site types depict rather well the productivity of the site under Finnish conditions where the terrain is flat and gently rolling, and where there are no real mountain ranges. The overburden left by the continental ice sheet is relatively thin and each soil class is covered by approximately the same types of plant community. In conditions where there are overlying soil strata of varying fertility, as well as mountain ranges, the site type is usually classified on the basis of tree age and height. Stand classification based on age and dominant height is used in Finland in thinning models. The forest site type system was adopted as the basis for determining land taxation classes in 1945.

The forest and peatland site type system has been of great value to research on the biodiversity of forest ecosystems and how to maintain it, as well as when evaluating the capacity of forest soils to withstand acidic deposition. Each forest site type has its own biotope characterised by a specific flora and fauna. It is possible, based on the areal distribution of the forest and peatland site types, to obtain a comprehensive picture of the extent and range of diversity.

The wood production capacity of forest site types was studied during the 1920's by preparing growth and yield tables for fully stocked pine, spruce and birch stands in a natural condition growing on different site types in southern Finland, and more recently corresponding tables for pine and spruce in northern Finland. The results showed that the forest site types depict the productivity of the site well enough for them to be used in planning wood production and in growing tree stands.

The first task of the Finnish Forest Research Institute (FFRI) was to establish long-term stand growth trials. While waiting for the results temporary sample plots were measured during the 1940's and 1950's in managed stands treated with cuttings, and growth and yield tables prepared for the main tree species. The growth studies gradually concentrated on the sown and planted stands which were becoming increasingly common in Finnish forests. We now have growth and yield tables, growth functions and thinning models for the dominant tree species and main types of mixed stand throughout the whole country.

The FFRI was given the task of inventorying the forest resources using statistical sampling methods. The field work for the 1st National Forest Inventory (NFI) was carried out during 1921-1924. When completed it was the first nation-wide inventory of its kind in the world. The forest resources have since been measured 8 times. The results form a 70-year-long time series of all the forests in Finland, as well as the major regions of the country. When forest parameters have to be estimated for small areas and individual forest holdings, the results are supplemented using aerial photographs and satellite imagery.

Wood utilisation and drain research was started during 1927-1933. As a result of this work, annual estimates of wood utilisation, removals and the stemwood drain were obtained by tree species and timber assortments for the forestry board districts and the whole country after 1950. The forest balance is obtained by comparing the drain with the NFI estimates of the growing stock volume and its annual increment. The forest balances have represented a reliable source of information about the development of the volume, structure and quality of the forest resources.

During the 1960's fears were voiced that the high level of wood consumption and expanding forest industry would result in overcutting and a reduction in the forest resources. This led to a research program lasting for almost 20 years into ways of increasing tree growth. Alternative silvicultural programs were drawn up covering improved cutting techniques, regeneration of under-productive stands, the drainage of peatlands, forest sowing and planting, fertilization and young stand tending. The effects of these measures on wood production and the economic viability of forestry were also estimated. The calculations included development prognoses for the growing stock, its increment and drain for periods lasting from 50 to 100 years.

The main results of the research was that, when intensive management is started, the removals can be immediately increased to a level above the increment without endangering the sustainability of production. The growing stock volume initially decreases, but the higher increment of the new stands will compensate for this reduction in the future. The increment and removals of the new forest are much greater than those obtained with earlier silvicultural treatments.

Broader and more comprehensive research is needed in conditions where the working and production techniques are rapidly changing. The mechanisation of forestry work and the new data procurement and interpretation techniques, such as aerial photographs, satellite imagery, automatic outputting and computers, require a

Fertility classes of forestry land based on the vegetation of mineral soil and peatland site types, and the range of potential overbark stemwood increment on these types in southern and northern Finland.

Fertility class	Mineral soils			Peatlands and heathy peatlands		
	Forest site type in southern Finland	Indicator plants	Annual increment, m^3/ha	Natural	Drained	Indicator plants
				Peatland and heathy peatland types		
1	OMaT Oxalis-Maianthemum type	Thread mosses, herbs and grasses, wood sorrel (Oxalis acetosella), may lily (Maianthemum bifolium), ferns (Filicinae), umbilleferous plants, mountain melick (Melica nutans)	8–3	Herbrich, herbrich and sedgerich spruce swamp and fenlike swamp	Herbrich heathy peatland	Ferns, smallreed (Calamagrostis purpurea), meadow-sweet (Filipendula ulmaria), marsh cinquefoil (Potenttila palustris), marsh marigold (Caltha palustris)
2	OMT Oxalis-Myrtillus type	Glittering feather moss, (Rhytidiadelphus), red-stemmed feather moss, wood crane's bill, other herbs, fingered sedge, grasses, blueberry	7–2.5			
3	MT Myrtillus type	Glittering feather moss, red-stemmed feather moss, herbs and grasses, blueberry, also lingonberry (Vaccinium vitis-idaea)	6–2	Myrtillus and shallow-peated spruce swamps, herbrich sedge pine swamp, ordinary sedge spruce swamp, herbrich sedge bog, fenlike pine swamp, eutrophic fen	Myrtillus heathy peatland	Myrtillus type herbs, sedges, bog-bean (Menyanthes trifoliata), marsch cinquefoil, blueberry, lingonberry, pine swamp dwarf shrubs
4	VT Vaccinium type	Red-stemmed feather moss, few herbs and grasses, lingonberry, also blueberry, heather (Calluna vulgaris)	5–1.5	Vitis-idaea spruce swamp, spruce-pine swamp, shallow-peated pine swamp and spruce swamp, sedgerich pine swamp, cotton-grass pine swamp, sedgerich open bog	Vitis-idaea heathy peatland	Blueberry, lingonberry, pine swamp dwarf shrubs, (Carex globularis), sedges, cotton grass (Eriophorum vaginatum)
5	CT Calluna type	Lichens, red-stemmed feather moss, very few herbs and grasses, heather and other dwarf shrubs	4–1.5	Dwarf shrub pine swamp, cotton-grass pine swamp, small-sedge pine swamp, Papillosum bog	Dwarf shrub heathy peatland	Pine swamp dwarf shrubs, cotton grass, few-flowered sedge (Carex pauciflora)
6	CIT Cladonia type	Lichens, some red-stemmed feather moss, scattered dwarf shrubs	3–1	Raised pine bog, fuscum pine swamp, small-sedge open bog, fuscum bog	Cladonia heathy peatland	Sphagnum moss, cotton grass, heather, lichens (Cladonia)

Finland is divided into forest vegetation zones extending from the south to the north: 1. Archipelago Finland, 2. southern Finland, 3. Ostrobothnia-Kainuu, 4. Southern Lapland, 5. Forest Lapland, 6. Fell Lapland. The forest and peatland site types listed above correspond to the situation in southern Finland: the other vegetation zones have their own parallel site types.

The vegetation cover of the Oxalis-Maianthemum forest site type (OMaT) is multi-storied, and there is considerable variation in the vegetation during the growing season. Wood anemone (Anemone nemorosa), hepatica (Hepatica triloba) and wood sorrel (Oxalis acetosella) flower in the spring and the ferns reach full size. Many herbs and grasses appear later on during the summer. There are no uniform moss and dwarf shrub layers, nor accumulation of undecomposed raw humus, because the plant and animal remains decompose rapidly. Highly fertile sites of this type are mainly found in southern Finland, and account for almost 1% of the forest land area. The photograph was taken in the Åland Islands.

The field layer of the Oxalis-Myrtillus forest site type (OMT) has an moss cover, and it contains demanding species such as feather mosses (Rhytidiadelphus). The indicator species used to classify the site include wood sorrel (Oxalis acetosella), fingered sedge (Carex digitata) and spring pea (Lathyrus vernus). The photograph shows an old planted spruce stand where the light passing through openings in the crown canopy has enriched the flora. Sites of this fertility class represent 11% of the forest land area in Finland.

Mosses and blueberry (Vaccinium myrtillus) form a uniform vegetation cover on the Myrtillus forest site type (MT). The most common moss species are red-stemmed feather moss (Pleurozium schreberi), glittering feather moss (Hylocomium splendens) and greater fork moss (Dicranum majus). Common vascular plants are wood cow-wheat (Melampyrum silvaticum) and hair-grass (Deschampsia flexuosa). The rate of decomposition of the plant litter is so slow in humid parts of central and northern Finland that a layer of raw humus, which retards tree growth, starts to form. Altogether 27 % of the forest land area belongs to this fertility class.

The field layer of the Vaccinium forest site type (VT) consists of a uniform cover of lingonberry (Vaccinium vitis-idaea), crowberry (Empetrum nigrum) and heather (Calluna vulgaris). The bottom layer comprises feather mosses. Lichens occur on rocks, stumps and hummocks. Vascular plants include common cow-wheat (Melampyrum pratense) and bush-grass (Calamagrostis epigeios). Compared to the classification of the ground vegetation carried out at the beginning of the century, there appears to have been an overall increase in site fertility. The main reason for this is probably the climatic warming that started in the 1930's, and nitrogen deposition. Vaccinium and corresponding forest site types account for 33 % of the forest land area in Finland. The photograph shows a permanent sample plot established for growth and yield studies. The trunks have been numbered and marked for the measurement of breast-height diameter.

The Calluna forest site type (CT) has a strong dwarf shrub layer dominated by heather (Calluna vulgaris), and there are a lot of lichens in the field layer. This gives the bottom vegetation a greyish appearance. Yellow bird's nest (Monotropa hypopitys), which flowers in August, can be seen in the foreground. The proportion of Calluna sites and those of corresponding fertility is 20 % of the forest land area.

The ground vegetation in the Cladonia forest site type (ClT) is almost exclusively composed of different species of lichen, without any herbs or grasses at all. This site type does not really occur in southern Finland. The proportion of Cladonia and corresponding site types is 9 % of the forest land area.

Spruce mires are peatlands with a tree cover of spruce or broad-leaved species. The peat layer is rather thin and the moving surface water contains oxygen. The photograph shows grey alder, pubescent birch and spruce growing on a wet, grassy spruce mire in a natural state. Its fertility makes it one of the best types of spruce mire. The ditching of spruce mires is a profitable undertaking. They account for one quarter of Finland's peatland area. The most fertile natural spruce mires are nowadays being protected as examples of fire refuges.

Ditching converts rather fertile spruce swamps into Myrtillus heathy peatlands, which are good sites for spruce. The photograph shows Jaakkoinsuo peatland at Vilppula in northern Häme, which was drained for the first time in 1909. The peat was originally 30 cm thick and the stand dominated by pubescens birch. A spruce understorey developed during 1930-45. The birch overstorey was felled in 1958, leaving a spruce understorey with a stem number of 1400 trees/ha and stand volume of 35 m³/ha. Altogether 74 m³/ha were removed in the thinnings made in 1958. The current stand volume is 300 m³/ha and the annual increment 11 m³/ha.

Drainage has converted this natural sedge pine mire into a pine-birch mixed stand. The site class is Myrtillus/Vaccinium heathy peatland, which corresponds to the Vaccinium forest site type on mineral soils.

Eutrophic fens have disappeared from southern Finland because they have been drained for agricultural purposes. Natural fens occur on calciferous soils in northern Finland, e.g. in Kuusamo, where the photograph was taken. The vegetation is lush. Eutrophic fens are left undrained.

THE HEIKINHEIMO TREE SPECIES TRIAL

Olli Heikinheimo, professor of silviculture, established a tree species trial at Vilppula in northern Häme in 1951. The plots with the main tree species, Scots pine, Norway spruce and European white birch, were established next to each other. The mixed forest had been clear cut the previous year prior to prescribed burning. The planting density was 2 500 seedlings/ha. The pine and birch seedlings were two years old and the spruce seedlings 4 years old at planting. The soil texture is sand and the site was classified as the Myrtillus forest site type. The photographs were taken 43 years after prescribed burning, and show how the individual tree species regulate the ground vegetation.

The leaf litter produced by birch prevents the development of mosses and dwarf shrubs, and nutrient cycling in the soil is faster and more effective than that in coniferous stands. The ground vegetation contains more herbs and grasses, e.g. the butterfly orchid (Platanthera bifolia) than that on a Myrtillus site with a coniferous tree cover.

Canopy closure in a pine stand results in the development of the feather moss and dwarf shrub layers, comprising blueberry (Vaccinium myrtillus) and lingonberry (V. vitis-idaea), typical of coniferous forest. The floristic composition corresponds to the Myrtillus forest site type.

The floristic composition of the vegetation in the spruce stand is rather poor owing to the low light intensity, abundance of mosses and slow decomposition of the humus.

considerable research and development input.

The threat posed to the forests by industrial and traffic pollutants has resulted in a number of comprehensive research programs on forest health and vitality, and on the causal factors involved. Structural changes in forest ownership and the drop in current removals below the increment level of the standing crop have increased research on the means available to forest policy. Multiple forest use and role played by the forest in the cultural landscape are being taken into account to an ever-increasing extent in the planning of wood production. The biodiversity of the forest and its dynamics under present conditions, as well as in a possibly changing climate, are the latest research topics.

Forest growing methods

Forestry changed from the mere removal of trees from natural forests to an activity centred around the growing of trees when the forest resources decreased and their quality deteriorated to such an extent that the sustainability of the wood supply was threatened. Forest management also presupposes that there is a demand for all assortments of timber. Rational growing and regeneration cuttings became an economically viable proposition when the pulp industry started to expand at the beginning of the 1950's. Prior to this, silvicultural research and experimentation had been used to investigate the principles underlying the growing of forest in accordance with the fire ecology laws of the boreal coniferous forest. The trees primarily removed in intermediate cuttings are those which, in natural forests, would otherwise die as a result of self-thinning. If the stand is not thinned, at least 25-30% of the production of stemwood will be lost through natural mortality before the stand has reached the final cutting stage.

The second aim of intermediate cuttings is to accelerate the growth of the final crop trees and to increase their value. Trees of inferior quality are removed in thinnings, as well as enough trees from the dominant crown canopy to give the trees of the final crop room to grow. The valuable final crop consists of high-quality, saw-timber trees. A large tree size and small size variation reduce the harvesting costs; the smaller the tree, the higher are the harvesting costs per cubic meter.

The third aim of intermediate thinnings is to achieve the sort of stand structure, at an age approaching final cutting, where the conditions are favourable for a natural stocking of seedling material. There are usually seedlings present below the trees at the final cutting stage on dry and dryish mineral soil sites and on wetlands. After final cutting a new seedling stand develops on the site. The formation of a natural seedling stand below spruce-dominated stands is promoted by the presence of birch.

When rationalised tree growing was first introduced there were many stands containing poor-quality trees and a stand density which was so low that the stand could not fully exploit the growth potential of the site. Cleaning cuttings were made in these stands, after which the trees were left to form a fully stocked stand before the next cutting.

Timber and income-producing intermediate cuttings are performed 2 - 5 times before final cutting, depending on the thinning model. The number of thinnings has decreased as a result of the mechanisation of harvesting. Between 30 and 50% of the stemwood crop is harvested in these thinnings.

The natural regeneration of spruce below broad-leaved species and pine is a part of normal succession in the sort of stand structure where cut-overs have been harvested from above the spruces gradually developing into the dominant stand. Birches are retained in even-aged spruce stands in order to maintain the health of the stand and the capacity of the site to produce seedlings.

A new seedling stand is established on the site after final cutting. Tree species which grow best on the sites in question are favoured. Suitable sites for pine are dry, infertile mineral soils and nutrient-deficient wetlands. Pine grows well on sites of the Myrtillus type, but on Oxalis-Myrtillus sites its growth is lower than that of spruce and it forms thick branches.

Spruce and birch grow relatively well on sites of the Myrtillus type and better. The most valuable crop of silver birch is obtained on fertile mineral soils. Pubescent birch grows better than silver birch on wetlands, and especially on drained peatlands where it improves the substrate for conifers. Aspen and grey alder grow best on fertile mineral soils.

A stand is ready for final cutting when the growth starts to decline and almost all the trees have reached the sawtimber size. Obtaining a favourable financial return presupposes that the trees are grown to sawtimber size even when they are used by the paper industry. This gives a high rotation yield/hectare/year, and a final crop in which the harvesting costs are as low as possible per cubic meter. The effect of tree size on the financial return shows that the harvesting costs of first thinnings per cubic meter are about three times the costs of harvesting the final crop.

Natural seedling material is best used for regeneration if it is available almost immediately after final cutting or if a viable seedling stock is already present under an old stand. Dry mineral soils, wetlands and drained peatlands readily produce natural seedlings. Sowing and planting are necessary if the tree species is to be changed or if natural seedlings have not appeared about ten years or more after final cutting.

Pine and birch are regenerated naturally using the seed-tree method. Pine requires about 30 - 50 seed trees/ha on dry mineral soils and on wetlands, and 50 - 100/ha on more fertile sites where the flourishing ground vegetation retards seedling stocking. On infertile, dry mineral soils there are usually so many pine seedlings under the old stand that the final crop can be harvested at the same time.

10 - 20 seed trees/ha are sufficient for the regeneration of birch. Soil preparation speeds up the formation of a seedling stand of all tree species. Unless seedlings are already present on the site before final cutting, soil preparation is needed to ensure a good stock of seedlings within a reasonable period of time.

The advantages of site preparation, and especially prescribed burning, lie in the fact that these techniques liberate the nutrients bound in the humus layer in a form available to plants and initiate the sort of succession that maintains biodiversity. Prescribed burning and site preparation reduce the risk of attack by insects and pathogens. Scarifying is nowadays the main type of site preparation. Mechanised timber harvesting breaks the ground surface and promotes the release of nutrients from the humus and the formation of a good stock of seedlings.

Spruce is regenerated by the shelterwood method. Seedling stocking is started by reducing the stand density in the final thinning down to a level where viable seedlings can develop under the old trees. 200 - 300 seed-producing shelter trees per hectare are required. The final crop trees prevent the ground vegetation from becoming too lush, and protect the spruce seedlings from direct sunlight and frost. If there are birches in the stand, these are left in the shelter stand.

Strip felling is clear-cutting in which seed are obtained from the border stands. The results are good on thin-peated spruce swamps where a seedling stand readily develops. The method is also used in the regeneration of spruce on mineral soils, where seedling stocking is slow without soil preparation. Owing to the slow rate of regeneration, planting often has to be employed in these areas.

Clear-cutting and regeneration by sowing or planting became a common practice during the 1960's when wood consumption threatened to exceed the sustainable removals. Sowing and planting are a part of effective, profitable forestry on the sort of sites where natural regeneration is slow or results in a low-value seedling stand. The final crop is obtained from a managed, planted stand 10 - 15 years earlier than that in a stand naturally regenerated from seedlings which had not appeared before final cutting. The results of forest tree breeding can be exploited in artificial regeneration. Windthrow losses and the other forms of damage that occur in seed and shelterwood stands can also be avoided.

The number of planted seedlings giving a good economic return is about 2 000 plants/ ha in the case of conifers, and about 1 700 plants/ha for birch. Natural seedlings supplement a planted seedling stand on sites given soil preparation. In a coniferous stand at the canopy closing, pole stage there should be about 2 000 trees, and in a birch stand about 1 000 trees. Owing to the vigorous natural regeneration of birch it is almost always present as an admixture in sown or planted stands, apart from on the most infertile sites.

If thin-branched, butt-logged quality pines are to be grown without pruning, there should be about 4 000 trees/ha, including other tree species, in a natural seedling stand. In evenly-spaced, similar-sized planted stands the branches at the base of the trunk are pruned naturally at a much lower stem number. The pruning of butt logs became a common practice during the 1980's. According to the growing model for large-sized, branchless butt logs there should be about 500 pruned trees/ha after thinning. In the final crop there will be 350 large-sized trees/ha.

Ground vegetation and sprouts which suppress the young trees are controlled in the early stages of seedling stand tending. Mechanical cleaning is used the most. In some cases chemical ground vegetation and sprout control is more effective and cheaper than mechanical treatment. When carried out correctly it does not harm the forest ecosystem. The use of chemical control has decreased owing to increasing concern for environmental values. During the past few years it has been used on about 2 000 ha a year, the area of mechanical seedling stand tending having been about 200 000 ha/year.

The seedling stand is tended in such a way as to maximise the number of timber-producing trees in the first thinnings. Even-aged pure stands are the most profitable. An admixture of broad-leaved species in coniferous stands is required to maintain stand health, nutrient cycling, diversity and scenic value. Forest damage is less of a problem in mixed stands than in pure stands.

In today's forests where there are no fires or slash-and-burn agriculture, a change of tree species is needed to maintain site fertility and forest health. Repeated growing of spruce on the same site increases the accumulation of raw humus, reduces soil fertility and increases damage caused by rot fungi. These problems can be prevented by replacing spruce stands with silver birch. One advantage of this is that spruce seedlings usually appear at no cost under birch, and are ready to form the next stand generation. Generation after generation of pine can be grown on dry, infertile sites without reducing site fertility.

Harvesting and regeneration methods have been developed for use in landscape forests and recreation forests close to inhabited areas which do not cause sudden, major changes in the scenery, and which retain the specific characteristics of the forests.

Development of the age structure of the forests and recommended stand rotation periods (final-cutting age).

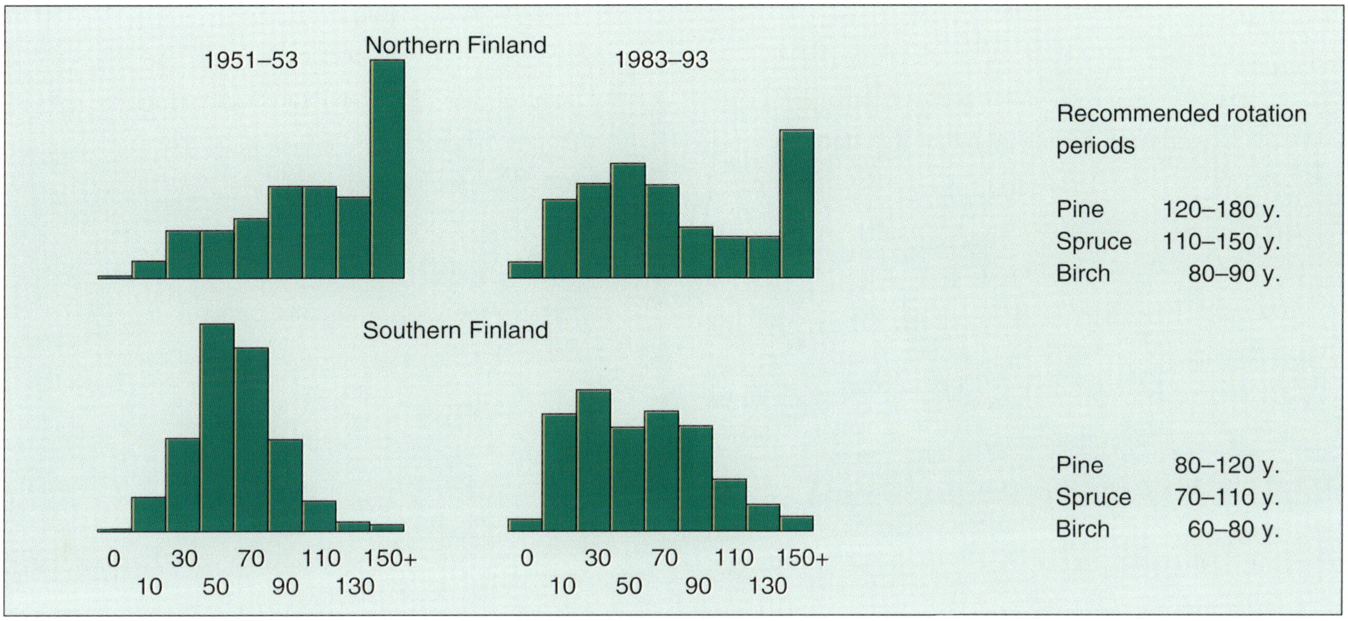

The recommended rotation period for a commercial spruce stand on the Myrtillus forest site type in southern Finland is 80 - 100 years. When the stand reaches the final cutting stage the volume increment is 3 - 3.5 % a year. The volume obtained in final cutting is 200 - 250 m3/ha, the proportion of sawtimber being 65 - 80 %.

Prescribed burning is a suitable silvicultural method on glacial till soils with a cover of spruce. Till soils are predominant in Finland. Burning an area of less than 1 ha is technically difficult and costly. A hand pump and radio telephone ensure that the fire does not get out of control.

Scarifying the burnt soil is necessary to obtain a vigorous stock of seedlings. The harrow is fitted with a seed-sowing device. The photograph was taken at Töysä in southern Ostrobothnia.

Scarifying in the spring ensures that the pine seed germinate in the mineral soil where they are not affected by competition from the other vegetation. Harrowing also promotes the natural regeneration of birch on this site at Töysä.

Three growing seasons after prescribed burning and sowing the site is covered with 15 to 25 cm-high pine seedlings at a stocking of at least 10 000 plants/ha. The seedlings in the Töysä regeneration area are still dwarfed by pioneer plant species, such as the fireweed (Epilobium angustifolium) shown in the photograph.

The pine stand artificially regenerated after prescribed burning has got off to a good start and the crown canopy is gradually closing. Seven years have passed since the burn.

The canopy closure stage has been reached in the pine stand sown after prescribed burning. Thinnings in the young stand have changed the tree species composition by favouring naturally regenerated silver birch (Betula pendula).

A planted pine stand becomes branchy on fertile sites of the Myrtillus forest site type at least. This is due to the site fertility and abundant growing space. The natural regeneration of pine is not usually successful on Myrtillus type sites. Branchiness reduces the saw timber quality. It can be improved by pruning the best trees in the stand, 500 trees/ha, when they are about 10 m high. The age of the pine stand is 25 years and the volume 120 m^3/ha. The stand is located on the coast of southern Ostrobothnia.

The use of seed trees in the natural regeneration of pine stands is the classical method on sites of the Vaccinium and less fertile site types. About 100 seed trees/ha have been left at Lieksa, eastern Finland. The soil has been harrowed to promote seedling stocking.

Seed trees are usually removed 5 - 10 years after establishment of the seedling stand. The photograph shows a pine stand on a Vaccinium type site at Lieksa in eastern Finland. The area was clear cut and harrowed in 1980. A dense seedling stand developed and the seed trees were removed in 1990. The photograph was taken in 1993 when the seedling stand was 1 m high.

Annosus root rot (Heterobasidion annosum) causes the greatest economic losses in Finland's forests. It occurs in old spruce stands that have passed the final cutting stage, and is most common in the Åland Islands, in southern and south-western Finland and along the Ostrobothnian coast.

Fertile land with a spruce cover can be replaced with planted silver birch (Betula pendula) after prescribed burning. Changing the tree species is a good practice if Annosum root rot (Heterobasidion annosum) is a problem in the area. The photograph shows a 2-year-old birch stand in Uusimaa Province in southern Finland.

*The same silver birch (*Betula pendula*) stand as that shown in the picture on page 159. It has survived moose damage and reached a height of 2 - 4 m in six years.*

*Early growth of silver birch (*Betula pendula*) is faster than that of conifers, and in favourable conditions the stand can reach the final cutting stage even within 40 years. The photograph shows a 28-year-old birch stand planted on an abandoned field near Jyväskylä in central Finland. The average yield has been 11 m^3/ha/year. Forest dominated by silver birch accounts for over 1% of the forest land area.*

The area regenerated annually by cultivation has varied from 110 000 to 150 000 ha during the past 30 years. This is equivalent to 0.55 - 0.75% of the forest land area. About 80% of the artificial regeneration is done by planting. After germination, the small seedlings are hardened in the nursery in peat balls.

Thinnings are essential in commercial forests in order to concentrate stand growth in the largest trees, to maintain nutrient cycling and to promote species richness. The spruce stand in the photograph was planted in 1932 on land of the Myrtillus forest site type at Alkkianvuori (altitude 190 m a.s.l.) near Parkano in southern Finland. This part of the stand has not been thinned. In 1987 the stem number was 1 650 stems/ha, basal area 40 m^2/ha, volume 365 m^3/ha, mean yield 6.6 m^3/ha/year, and mortality 10% of the stem number. The photograph was taken in 1993.

This part of the Alkkianvuori spruce plantation has been thinned twice, the removal totalling 90 m^3/ha. In 1987 the stem number was 500 stems/ha, basal area 20 m^2/ha, volume 190 m^3/ha and mean yield 5.2 m^3/ha/year.

THE HEIKINHEIMO SLASH-AND-BURN EXPERIMENT

Olli Heikinheimo, professor in silviculture, established an experiment on land owned by the Finnish Forest Research Institute at Vesijako, near Evo in Häme: pine seed were sown among the rye sprouts in the slash-and-burn area in spring 1918. A large number of seedlings developed, the seedling density being 400 000 plants/ha three years after sowing. Natural birch seedlings also appeared in the area. The fertility class is Myrtillus forest site type on stony soil. About 80 % of the seedlings were removed in a thinning. Two plots were marked out in 1948; the trees on one plot were left to grow untreated as a control, and on the other the pine stand was managed as commercial forest. The density of the pine stand when the experiment was established was about 15 000 trees/ha and the dominant height 12 m. The stands in the photograph are 74 years old.

The effects of thinning have been investigated in an extensive series of experiments by the Finnish Forest Research Institute. Compared to unthinned stands, thinning and reducing the stand density to encourage natural regeneration reduces the total yield of pine and spruce stands by 10 - 15 % during a normal rotation period on sites of medium fertility. Non-thinning means a 20 - 30 % increase in mortality during the stand's lifetime. Although reducing the standing crop capital by means of thinnings decreases the growth, the expected value of the stand is increased because cutting income is obtained at an earlier stage, and the stumpage price of the larger stems is raised.

The total yield of an unmanaged pine stand during a 68-year-period has averaged 7.2 m^3/ha/year, the share of mortality being 2.2 m^3/ha/year, i.e. 30 % of the yield. The stand volume in 1986 was 383 m^3/ha and the stem number 3 640/ha.

This commercially managed pine stand has been thinned four times. The total yield of the stand during a 68-year-period has averaged 7.4 m^3/ha/year. The average yield obtained in thinnings has been 2.9 m^3/ha/year, i.e. 39 % of the yield. The stand volume in 1986 was 303 m^3/ha and the stem number 550/ha.

Forest soil ploughing has been used as a site preparation method on till soils in northern Finland since the 1960's. Ploughing increases the soil temperature, and improves nutrient and water availability on the shoulder of the furrow. Old spruce stands which do not readily regenerate naturally have been ploughed and planted with pine. Ploughing is currently being replaced by scalping and hummocking.

Despite the rather harsh climate in the region, vigorous young pine stands have been achieved in the ploughed areas. The photograph was taken on the Arctic Circle and the stand was planted 25 years earlier. The stand volume is about 100 m^3/ha.

In 1973 lightning struck a spruce in the Oulanka National Park in northern Finland. The tree fell, caught fire and started a forest fire. About 1 ha of thick-mossed, Hylocomium-Myrtillus spruce forest was destroyed. The photograph was taken 19 years after the fire. Although the dwarf shrub and moss vegetation was destroyed, they have already returned. Pioneertree species have just started to appear.

The Oulanka National Park is bordered by the Kuusamo collectively owned, private forest district where commercial forestry is practised. The tree stand in the clear-cut area was felled about 10 years ago and planted with pine. An old thick-mossed spruce stand in the national park, partly destroyed by a small forest fire in 1973, can be seen in the background.

The growing of uneven-aged stands in Finland has not been very successful. Spruce is a semi-shade tree and it is the only tree species capable of regenerating naturally under a shading tree stand in Finland's climate. Growing repeated generations of spruce on the same site deteriorates the fertility of the site and prevents seedling stocking. The photograph was taken in the Liesjärvi National Park in southern Finland where the thick moss layer and lack of light inhibits seedling development.

The growing of uneven-aged stands is biologically possible under the conditions prevailing in central Europe. Silver fir (Abies alba) and beech (Fagus silvatica) are shade trees, i.e. they regenerate naturally in the shade under a tree stand. Higher temperatures and fertility promote the decomposition of the plant litter. The photograph was taken in the Black Forest in southern Germany.

FOREST IMPROVEMENT

The results of the 1st National Forest Inventory and growth and yield research carried out during the 1920's showed that the growth of the standing crop could be approximately doubled by restoring the extensive areas of forest ruined by selection felling, and by regenerating the under-productive and over-aged stands.

Forest improvement work initially became common in the forests owned by the state and by companies. Because it takes at least 20 years to obtain a return, in the form of income from wood sales, on most basic improvement investments, private forest owners were rather reluctant to make such investments without state support. The Forest Improvement Act was passed for prescribed periods in 1928, and as a permanent act in 1967. According to the act, state funds are provided in the form of grants and long-term, low-interest loans for long-term forest improvements, in addition to the forest owners' own funding. This system has greatly increased the forest owners' interest in silviculture and taught them to realise that carefully planned forest improvement can also be profitable as private investments.

The national benefits of forest improvement are an increase in work and income in areas with high relatively high unemployment, and an increase in the production and exports of the forest industries and allied sectors. Forest improvement increases the sustainable removals immediately, because the subsequent increase in growth compensates for possible initial reductions in the standing stock.

The ditching of wetlands and waterlogged soils, which started in the 1930's and reached a peak in 1970, has had the greatest effect on the growth in the standing crop. The annual increase in increment achieved through ditching is about 11 million m3. Fertilization is also required on nutrient-deficient wetlands.

Extensive regeneration of low-productive forest started at the end of the 1940's. Cuttings for firewood and industrial wood were directed at the regeneration of stands ruined by selection thinnings, spruce stands growing on dry mineral soils and low-value broad-leaved stands. The regeneration of low-productive stands resulted in sowing and planting becoming common practices.

Other forms of forest improvement work include seedling stand tending, the fertilization of middle-aged stands, branch pruning and the construction of forest roads. Forest roads have greatly reduced the costs of harvesting and transportation, as well as silvicultural costs. They have also promoted recreation use of the forests and the picking of wild berries and mushrooms.

Forest improvement grants are available for planning work and to cover the costs of seeds, transplants, fertilizers, equipment and supervision. A forest improvement loan is not covered by an inflation index, and interest-free periods have ranged from 1 - 10 years depending on the length of the loan and regional policy. The interest on long-term loans has been 3 %, and on short-term ones 5 %. Temporary tax exemption on area-based forest income has been granted for improvement areas.

The investments increased during the 1960's and reached a peak during 1973-1975 when Finland received a forest improvement loan of USD 20 million from the World Bank.

The ditching of natural peatlands has almost completely ceased in Finland, and nowadays the main emphasis is placed on maintenance ditching. Sphagnum mosses choke up the ditches if they are not cleaned regularly. The photograph shows a second generation stand on a heathy peatland where Sphagnum peat is spreading as the groundwater table rises.

Forest roads are primarily built for transporting timber, but they also benefit all other forms of forest use. They also serve as ecocorridors.

Timber logging and transport

During the days of manual labour the trees were felled and processed by axe, handsaw and barking iron, hauled by horse and sledge to the edge of the floating routes, and then floated down the rivers to the consumer. Horses were also used in long-distance transport. At the beginning of this century rail and road transportation started to increase. Market wood and most of that used on the homesteads were felled and hauled over the snow and ice roads during the winter when the watercourses and wetlands were frozen.

Forestry work was made even heavier by having to live in temporary camps, which the workers had to arrange for themselves as well as their food. Legislation was eventually introduced making meals and accommodation the responsibility of the contractor. Mechanisation, trucks and the development of the forest road network made it possible for the forest workers to travel to work each day from their homes. The cabs of logging machines afford protection against the weather.

The manual period lasted up until the beginning of the 1960's. At that time the number of men employed annually in forestry work ranged, depending on the economic situation, 110 000 to 140 000. The seasonal nature of the work is illustrated by the fact that the monthly variation in the work force was 60 000 - 186 000 still in 1964. There were 60 000 horses engaged in forest work, and they performed 15 000 horse years in 1951.

War and reconstruction slowed down mechanisation compared to the other countries in the boreal coniferous forest zone. The agricultural population was large and was in fact increased by the establishment of new holdings after the war to settle the displaced farmers from the ceded territories, and landless farm workers. The mechanisation of forest work was intentionally slowed down in order to maintain employment. The emphasis in forestry rationalisation work was placed on the development of cutting work performed manually with a chainsaw, and forwarding by tractor.

The mechanisation of agriculture and the movement of the farming population into other trades and professions occurred at a fast pace during the 1960's. The rise in the general wage level also increased the costs of forest work. The use of horses in haulage decreased. Work rationalisation and mechanisation, and the training of the labour force, started in the forestry sector.

The axe and handsaw disappeared from use during the 1970's. Debarking of the wood took place at the mills. Most of the wood was cut by forest workers with a chainsaw up until the 1980's. Farm tractors were used to some extent in forest hauling, but they proved to be too unsuitable for the forest terrain and thick snow cover. The numbers of forest tractors, multi-process machines processing timber after felling, and harvesters performing felling, debranching, cutting and piling increased.

In 1990, harvesters were used for 50 % of the final cuttings and 15 % of the thinnings, and forest tractors performed 90 % of the forest hauling. The share of fully mechanised timber extraction is expected to reach 90 - 95 % by the end of the century.

Farm tractors fitted with appropriate equipment are used in delivery logging in farm forests. The proportion of delivery wood sold at the roadside is about 25 - 30 % of all the wood harvested. Most of the wood in private forests is sold by stumpage sale. Contractors working for the buyers perform the harvesting and forest hauling.

During the 1980's the rationalisation of forest work

The productivity of timber logging and transport is high at final-cutting sites because the stemwood and outturn are large per areal unit. The mean area of stands marked for final-cutting in privately owned forest in southern Finland is 2 ha, and in the north somewhat larger. In stumpage sales the amount of timber cut in marked stands averages 500 m³; in stumpage sales the timber buyer is responsible for carrying out the felling and forwarding work.

increased productivity by an annual rate of 4 % in the timber processing, and by 9 % in forest hauling. The use of efficient working methods has resulted in higher stumpage prices for wood than in the other countries in the boreal coniferous forest zone.

Use of the Scandinavian assortment method in harvesting has also proved to be competitive in other countries. It has been made possible through the use of wheeled and tracked tractors suitable for different types of terrain and working conditions. The machines used in final cuttings have to be heavy owing to the great weight of the trees, but their construction is such that soil compaction is minimised and erosion prevented. They break up the humus layer and thus promote nutrient mobilisation and the formation of seedling stocks. Light, easily manoeuvrable machines, which do not damage the standing trees, are used in thinnings. Harvesting on soft, boggy land is done as far as possible when the ground is frozen and covered in snow.

In 1950, 65 % of the timber transportation was done by floating, 20 % by rail and 15 % by road. In 1991 the corresponding proportions were 15, 10 and 75 %. Nowadays floating is only practised in the form of bundle floating along lake routes. The last loose-log floating was carried out on the River Kemijoki in 1991. The advantage of road transport is the fast delivery time. The wood is delivered green and in prime condition to the mills, and the interest costs of storage are reduced. If the price of energy increases from its present level and if emission controls are set on road traffic, the competitiveness of floating and rail transport will exceed that of road transport.

In Scandinavia wood is harvested using the assortment method, the trunks being cut into the required lengths in the forest. The main advantage of doing this in the forest is that the timber can be transported out of thinned stands without damaging the standing trees.

The harvesting conditions during the short winter days are difficult. However, this is somewhat offset by the advantage offered by frozen ground in soft, boggy terrain when heavy loads of wood are being moved out of the forest by forwarder. Carrying out thinnings in spruce stands during the winter and the use of 60 to 70 cm-wide tractor wheels reduces soil compaction and damage to the trees left standing in the forest.

Annosus root rot (Heterobasidion annosum) is the most serious damaging agent in spruce stands in southern Finland. It spreads through the cut surfaces of stumps left by timber harvesting during the summer. The fungus can be controlled by spreading a competitive rot fungi on the stumps before Annosus root rot has had time to infect them. Artificial stump infection with Phlebiopsis gigantea prevents the appearance of Annosus root rot . A water suspension of the spores of Phlebiopsis gigantea is sprayed onto the stump surface when the air temperature is above +8°C.

In cuttings done by harvester, timber measurement, cutting into lengths and protective treatment of the stumps are all automated processes.

In harvester cutting the butts of rotten trees can be left in the forest as a nesting site for birds. The forest owner or company timber buyer marks which trees are to be left when marking the stand.

The forwarder driver spends his working day of up to 10 - 12 hours in the tractor. In the latest Timberjack models special attention has been paid to the ergonomic properties of the cabin, work safety and automatic controls. Contractor Matti Mäkelä is outputting his day's results from a computer. On the average, 120 - 150 m3 of timber are processed and transported to the side of the forest road during one day's work at a final cutting site.

Forestry planning

A forestry plan includes standwise information about site quality and the tree species composition, age, stem number, growth and timber assortments of the tree stand. These data are used to calculate a 10-year compartment-wise program for cuttings, silvicultural and forest improvement measures and other work. The sustainability of the removals is analysed using tree stand development estimates covering a number of decades. In addition to qualified experts, the forest owners and those responsible for carrying out the work participate in the planning work.

Forestry planning started in the state-owned forests and major private estates during the 19th century. The state and company-owned forests were included in the periodic inventory and planning programs carried out during the first few decades of this century.

Management plans for farm forest gradually became more common after the forestry boards started to make state-subsidised regional planning inventories of all the privately-owned forest land. Management plans for individual holdings are prepared from these inventories as a chargeable service. The field work was carried out at an annual rate of about 1 million ha during the 1980's. They inventories now cover about 75% of the privately-owned forest area. The regional planning results are used to promote forestry over extensive areas and in investigating the timber supply prospects.

National and regional forestry planning has been based on the results of the national forest inventories, as well as on wood consumption and removal statistics. The national plan up to the 1980's drew up estimates of the maximum sustainable cutting potential for the development of the forest industry, and alternative input programs for increasing wood production. At a time when the removals remain permanently below the increment of the standing stock, there is an increasing need to determine which measures would ensure total exploitation of the wood production potential and the retention of biodiversity in the forests.

The most important national programs commissioned by the government or instigated by the forest sector are:

- Program of the Forestry Planning Committee, 1961
- Report of the Agricultural Committee, which includes, as the target of the forestry intensification program, increasing farm forestry income in the face of agricultural overproduction, 1962, 1964
- Forestry Financing Programs, Mera I 1965, Mera II 1966, and Mera III 1969
- The Forestry Programs on Alternative Intensity Levels for the Economic Council. 1969
- Forest 2000 Program for the Economic Council, 1985
- Revised Forest 2000 Program, 1992

The main result of the planning programs made during the 1950's and 1960's was that, by improving cutting methods, intensifying silviculture and increasing forest improvement work, the sustainable removals can be immediately increased to a level above the increment in the standing stock without endangering the sustainability of production. This is due to the fact that the initial reduction in the standing stock will be compensated by the increasing increment of the new managed stands. The results of repeated national forest inventories show that this is now the case. The increment of the nation's forests has increased more than would have been expected on the basis of the planning calculations.

Changes in the expectations set on the forest and forestry

also change the scope and contents of planning. In addition to wood production, forest management also strives to maintain environmental benefits and biodiversity. The aim of planning has been to balance out the contradictory values guiding forest management.

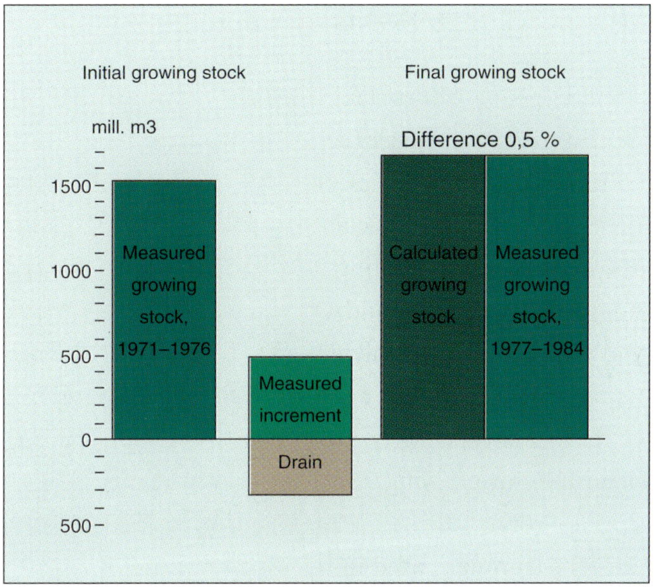

Forest balance for the period between the 6th and 7th National Forest Inventories. Calculated growing stock is initial stock + increment - drain. Drain has been obtained from research on the annual wood consumption and removals. A difference of only 0.5 % between the calculated and measured final growing stock demonstrates the reliability of the forest balance. Source: Kuusela et al., 1991.

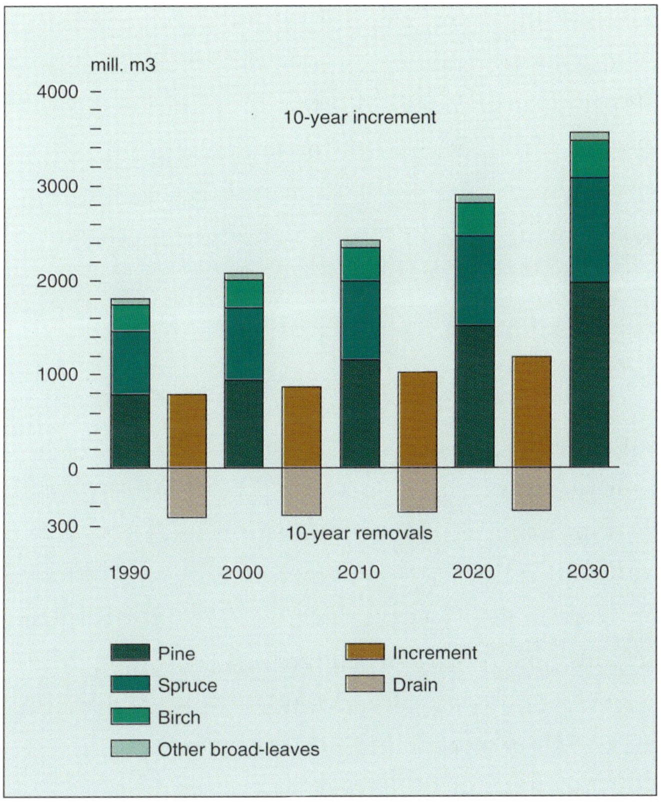

Development of the growing stock, its distribution into different tree species, and increment in alternative 3 of the Forest 2000 Program, in which the removals from production forests remain at the same level as in 1985-1989.

The achievements of progressive forestry

Improved cutting techniques and silvicultural methods, the regeneration of low-productive stands, the ditching and fertilization of peatlands, have all increased the area of forest land and the increment of the standing stock, and directed the forest resources along a course of continuous growth in the future.

According to the results of the national forest inventories, the volume of the standing stock decreased slightly at the beginning of the 1960's and then started to increase. The increment estimate for the standing stock initially remained constant and then started to increase as a result of forest improvement and the increase in the growth of new forests.

The magnitude of the removals, despite the harsh climatic conditions in Finland, is demonstrated by the fact that during the period 1952 - 1990 the drain from the forests was 1.33 times as high as the volume of the whole standing stock in 1952. Despite this, the volume of the standing stock increased by 22 % and the increment of the standing stock by 44 %.

The proportions of valuable conifers and sawlogs have increased. The felling value of the standing stock increased at a rate of FIM 1 200 million per year (1992 monetary value) during the period 1952 - 1990. The increase in value per year is almost as great as the mean total sum of silvicultural costs and long-term basic improvement investments of FIM 1 300 million/year during 1963 - 1992. At the same time the mean annual gross stumpage income was FIM 7 700 million in 1992 monetary value.

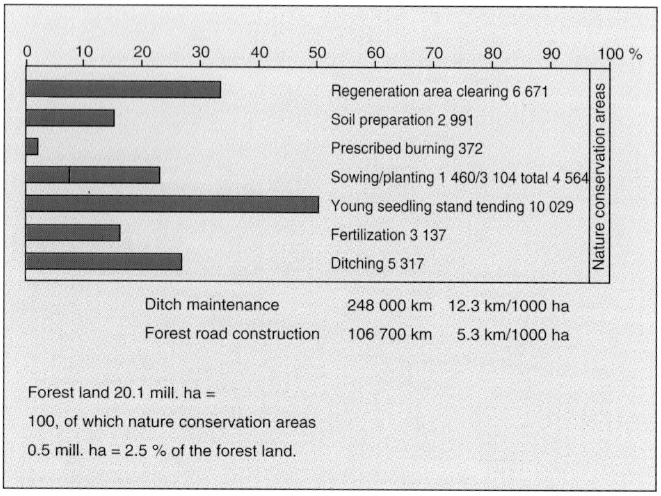

Silvicultural and forest improvement performances (1000 ha) during 1950-1992 with respect to the total forest land area.

Increment, 1952	55.0 mill. m3/y.
Growth increase resulting from drainage carried out prior to 1952	+1.5 mill. m3/y.
Growth in 1990 without inputs if removals had been as large as during 1952–1990	49.0 mill. m3/y.
Drainage and fertilization of peatlands	+11.0 mill. m3/y.
Changeover from selection cutting to silvicultural thinnings and regeneration of old and underproductive stands, sowing and planting, results of tree breeding and seedling stand tending	+17.0 mill. m3/y.
Fertilization of middle-aged stands	+2.0 mill. m3/y.
Increment in 1990	79.0 mill. m3/y.

Effect of different factors which have increased the increment of stem wood.

Crowing stock in Finland in 1951-53 and 1990.

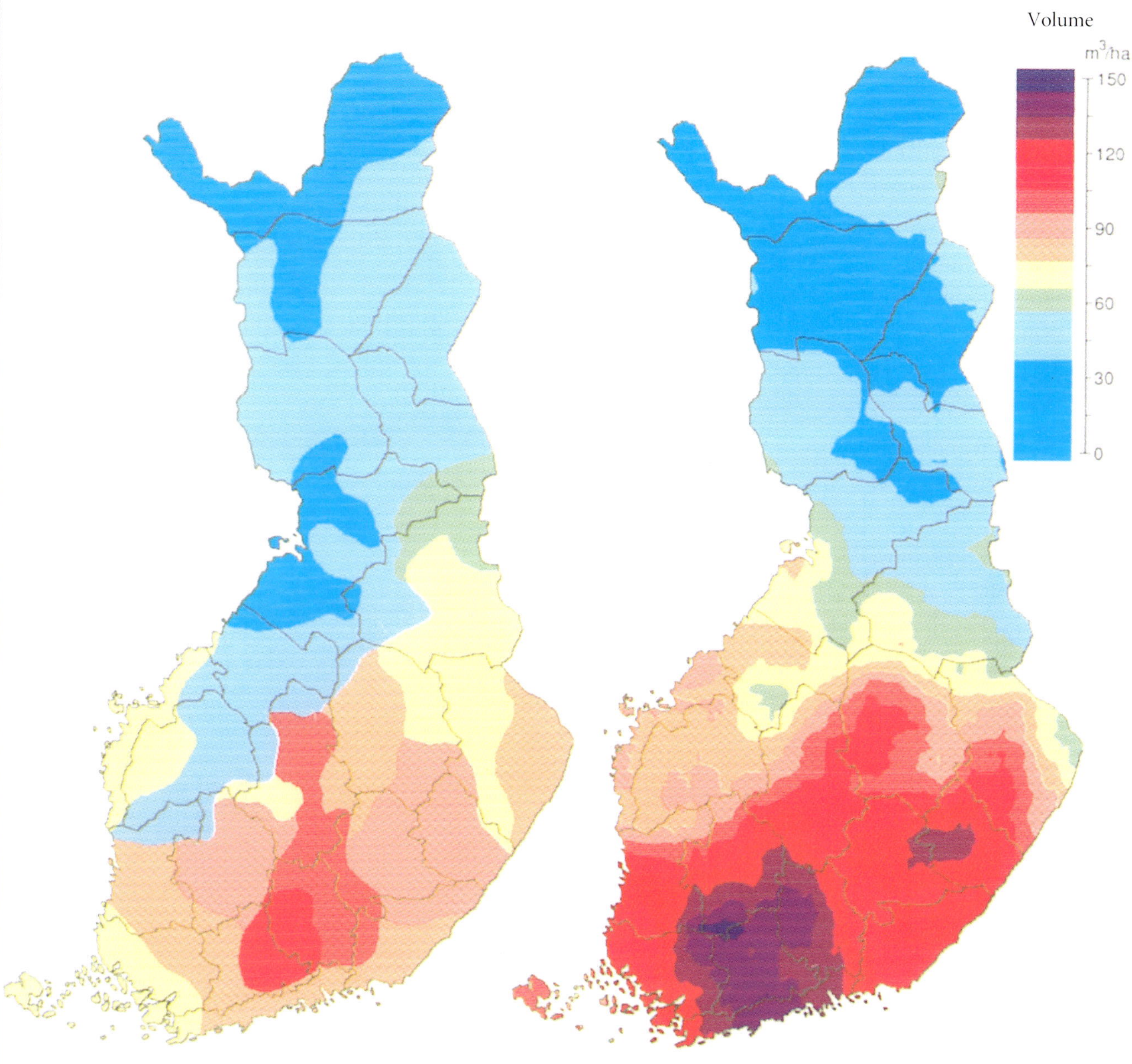

Mean volume of the growing stock during 1951-53 on forest and scrub land.

Mean volume of the growing stock in 1990 on forest and scrub land.

MULTIPLE FOREST USE

The forest has always been a multipurpose natural resource and environment for the Finns. There was little conflict between wood production and other forms of use for as long as wood was an essential commodity satisfying man's everyday needs. The idea of multiple forest use and the integration of wood production with other forms of use first came to light at a time when the proportion of workers engaged in agriculture and forestry out of the total work force fell below 20 %, and the diversification of employment and high standard of living resulted in forestry and its significance for the national economy becoming alien concepts to the majority of the population.

The fact that many people consider wood production, and especially the marks left by forest regeneration, to be a hindrance other forms of forest use is a social problem of the post-industrial, high standard of living. It is the case even in Finland, where there are 4.0 ha of forest land per capita, 5.7 ha of forestry land and 0.5 ha of conserved areas outside the sphere of commercial forestry. In most European countries the forest density is only 0.1 - 0.2 ha per capita.

The picking of wild berries and mushrooms, hunting and outdoor recreation are the most popular forms of multiple forest use. In the Nordic countries, berry and mushroom picking and outdoor recreation fall within the public right of access to all areas of forested land, apart from in the immediate vicinity of inhabited buildings. Damaging the forest is prohibited. Lighting open fires requires the landowner's permission. Hunting rights are tied to land ownership. Inhabitants of districts in the far north can hunt on state-owned land without payment. The state grants and sells permits to hunt moose and deer. The hunting of large predators is controlled.

Reindeer husbandry is practised in the reindeer herding areas that are mainly located in northern Finland. It is the main source of livelihood for the Lapps living in the three northernmost districts, and provides significant additional income to other people in northern Finland.

Rural settlements and the road network, forest roads and 300 000 country cottages and villas make multiple use forestry possible in all parts of the country. The most isolated wilderness areas can be reached by car and a relatively short walk. The park forests around the cities and other settled areas, the national parks and recreation forests on state land and the winter sport centres, are all well frequented sites for multiple forest use.

Wild berries	45.7 mill. mk	Value of game crop	
Wild mushrooms	6.0 mill. mk		
		Forest fowl	24.0 mill. mk
Total	51.7 mill. mk	Field fowl	4.5 mill. mk
		Water fowl	39.7 mill. mk
		Hares	19.7 mill. mk
Exports of ornamental lichen	11.0 mill. mk	Fur animals	16.8 mill. mk
		Moose and deer	140.1 mill. mk
Reindeer husbandry	80.0 mill. mk	Total	244.4 mill. mk

Income from forest products other than timber, 1992.

Conservation areas and land classes		Northern Finland	Southern Finland	Total	
		%			1000 ha
National parks		24.4	1.9	26.3	
Nature parks		5.6	0.2	5.8	
Peatland conservation		14.8	1.0	15.8	
Herb-rich forest conservation		0.0	0.0	0.1	
Special areas		1.1	0.9	2.0	
Wilderness areas		50.0	–	50.0	
Total	%	95.9	4.1	100.0	
Total	1000 ha	2 460.0	103.0		2 563.0
Conservation areas out of forestry land	1000 ha	13 810.0	12 544.0		26 354.0
	%	17.8	0.8	9.7	
Of forest land	%	4.8	0.3	2.2	20 074.0
Of scrub land	%	26.3	1.8	21.7	3 112.0
Of waste land	%	50.5	9.1	45.3	3 040.0
Roads, landings etc.					128.0

Nature conservation areas in 1993 and their share of the different forestry land classes.

Lingonberry (Vaccinium vitis-idaea) is commercially the most important forest berry, and the picking income accounts for more than one half of the total income from berry and mushroom picking. The size of the crops varies considerably from year to year. Berry-picking income in Finland is non-taxable, and is important supplementary income for people living in isolated parts of northern Finland.

False morel (Gyromitra esculenta) is a species which benefits from soil preparation in regeneration areas. It appears one or two years after scarification, often at the same time as the pine germlings. The fungus thrives on sandy soil and it can produce large crops.

Cloudberry (Rubus chamaemorus), which grows on spruce and pine swamps, is a valuable species. The size of the crop is highly dependent on the weather during the flowering period. When a natural mire is converted by drainage into a heathy peatland, the cloudberry crop often decreases. On the other hand, a lot of cloudberries are often picked along the banks of the ditches.

THE FOREST IN A CHANGING ENVIRONMENT

The tree species and other plants in the boreal coniferous forests have become adapted to major environmental changes. They have the genetic capacity to withstand extremes of climate. The acidic deposition falling on the forests and the increase in greenhouse gases in the atmosphere can change the environmental factors in ways which are not yet known.

The use of fossil fuels, such as coal, anthracite and oil, as a source of energy is the prime factor modifying the environment. These deposits, as well as the peat that has been formed under anaerobic conditions over extended periods of time, represent a huge store of carbon. So much carbon dioxide and mineral material are released into the atmosphere when they are burnt that the natural factors which normally regulate the material composition of the biocenosis, are no longer able to maintain a balance.

The sulphur and nitrogen oxides emitted from pollution sources are carried for hundreds and thousands of kilometres by the wind and air masses and eventually deposited on the forest. Direct fume emissions damage the needles and leaves in the immediate vicinity of the pollution sources. At abnormally high levels they have killed the trees and ground vegetation around mineral and metallurgical plants.

The most serious effect of acidic deposition is considered to be the changes it causes in the nutrient status of the soil. An increase in soil acidity accelerates nutrient leaching and the dissolution of iron and aluminium compounds into the soil water. Some of the dissolved metals are toxic to plants.

About one third of the pollutants deposited in Finland are derived from the country's own industry and traffic. The rest is carried in by the wind and rain from the south-west, south and south-east, and to a lesser extent from the east.

Total annual sulphur deposition according to the deposition model for 1990 was 7 - 10 kg/ha in the southernmost parts of the country, 4 - 5 kg/ha in central and northern Finland, 3 kg/ha in western Lapland and 5 - 7 kg/ha in areas along the eastern border in northern Finland. The critical sulphur load for the most sensitive forest ecosystems is considered to be about 7 kg/ha/year. The possible harmful effects of acidic deposition on the nutrient status of the soil are currently being investigated. Not enough is known about the capacity of the naturally acidic soils in the boreal coniferous forest zone to buffer acidic deposition.

Total annual nitrogen deposition is about 10 kg/ha along the southern coast and in south-west Finland and 6 - 2 kg/ha to the north of this region. The critical nitrogen load is considered to be 10 - 20 kg/ha/year.

The trees are damaged or killed to some extent in the immediate vicinity of the emission sources. The overall growth of the standing crop has increased more than expected, in part due to the fertilizing effect of nitrogen deposition and the increase in atmospheric carbon dioxide concentrations.

It has been widely predicted in Finland, as well as in central Europe, that the long-range transport of pollutants will kill the trees over extensive areas. However, these predictions have so far not come true. Tree defoliation, i.e. the loss of needles or leaves, is the main parameter used to depict the state of health of the forests. However, the effect of other factors causing defoliation has not been taken sufficiently into account. Other factors include the self-thinning of trees in dense stands, high tree age, high seed crops, weather affecting the trees' preparation for winter, cold, desiccating wind, severe frost together with a thin snow cover, and insect and fungal damage.

The reduction in the abundance of lichens growing on tree branches and trunks and other changes in the forest flora are evidence of the effects of deposition. According to field experiments involving the artificial raising of soil acidity, acidic deposition will eventually weaken the nutrient status of the soil.

The reduction of sulphur and heavy metal emissions in Finland and other western European countries has been so great that the damaged forests have started to recover in areas earlier subjected to high deposition levels. However, nitrogen deposition levels have continued to remain high. According to nitrogen fertilization experiments carried out in Finland, it is evident that current nitrogen deposition levels will not have any adverse effects on forest health in the near future.

The increase in the concentration of carbon dioxide and other greenhouse gases in the atmosphere is considered to be the greatest emission-derived threat to the environment. The higher the concentration of these gases, the greater the amount of outgoing heat radiation that is retained and the higher the air temperature.

The carbon dioxide concentration of the atmosphere was 280 ppm in 1800, 314 ppm in 1950, and 350 ppm in 1990. If the production of energy from fossil fuels continues at its present level, emissions of greenhouse gases will double by the year 2030. It has been estimated that this will result in a 1 - 3°C increase in the average global air temperature by 2030. The regional distribution of rainfall will change, and large parts of the present agricultural regions will be turned into desert.

According to predictions, the air temperatures close to the poles will rise the most. The temperature in Finland during the growing season could be 2 - 3°C higher, and during the winter 4 - 6°C higher, during the first half of the 2100's. Precipitation will also increase.If these prediction become a reality, broad-leaved trees will become more competitive than conifers. The natural coniferous forest zone will move to the north. Damage caused by insects, fungi and the wind will increase. However, silviculture will be able to maintain conifer-dominated forest in the same way as it is currently doing in those parts of Central Europe where beech and oak are the natural dominant trees. The costs of silviculture will increase, but this will be partly compensated by greater wood growth and yield.

INDUSTRIAL EMISSIONS ARE DESTROYING CONIFEROUS FOREST IN THE KOLA PENINSULA

The metallurgical industry in the Kola Peninsula has been polluting the forests and watercourses ever since the 1930's. The smelters are located in two large industrial complexes, one at Monchegorsk and the other at Nikel-Zapoljarnyi. The town of Nikel lies on the banks of the Paatsjoki River on the Russian-Norwegian border, 40 km from the Finnish border. Monchegorsk is located about 120 km to the east of Finland.

Although Russian naturalists started their own research already during the 1970's, researchers in the Nordic countries were not aware of this. Joint research did not start until the latter half of the 1980's. In 1989 the Finnish Forest Research Institute initiated the Lapland Forest Damage Project as a joint undertaking with scientists from the Kola Science Centre, the aim of the project being to determine the effects of pollutants from the Kola Peninsula on the vitality and state of health of the forests in the region.

According to the Nature Conservation Committee of the Murmansk region, in 1990 sulphur emissions from Monchegorsk amounted to 233 000 tonnes, from Nikel 190 000 t and from Zapoljarnyi 67 000 t. Nickel emissions from Monchegorsk were 3 000 t, and copper emissions 2 000 t. The damaged area around Nikel was about 25 000 ha, and around Monchegorsk 85 000 ha. The effects of these emissions are visible in forest ecosystems up to a distance of 50 km to the west of the industrial complex.

Tree death close to the smelters appears to support the hypothesis that heavy metals increase the leaching of plant nutrients from the soil, resulting in severe metabolic disturbances in the trees. This reduces the frost resistance of the trees and kills those parts of the trees located above the snow cover line. The effects of deposition have not been detected in the growth and state of health of the forests on the Finnish side of the border.

The vegetation is almost completely destroyed in the immediate vicinity of the emission sources, and the area has become an industrial desert ravaged by erosion. The annual mean sulphur dioxide concentration in the air in this area is 200 μg/m³.

*The most resistant dwarf shrubs are bearberry (*Arctostaphylos uva-ursi*), lingonberry (*Vaccinium vitis-idaea*) and crowberry (*Empetrum nigrum*). The branches and crowns of the spruces are dead above the snow cover line.*

Birch and willow are more resistant to the emissions than conifers. They survive in conditions where the rest of the vegetation has been destroyed.

Glittering feather moss (*Hylocomium splendens*), which is the indicator species for thick-mossed spruce stands (*Hylocomium-Myrtillus* forest site type), has completely disappeared from the field layer and been replaced by crowberry (*Empetrum nigrum*). Crowberry is more resistant to pollutants than other dwarf shrub species. The famous Swedish botanist Carl Linné was already aware of this when he reported that only crowberry could survive the sulphurous gases surrounding the Falun copper mine. This spruce stand is dying about 25 km away from the emission sources in Monchegorsk.

This thick-mossed spruce stand (*Hylocomium-Myrtillus* forest site type) is located in eastern Lapland, on the Finnish side of the border, under the same sort of climatic conditions as the spruce stand near Monchegorsk. As there are no local emissions, there is an intact, uniform moss stand.

The deposition of pollutants from the metallurgical industry have produced an industrial desert around Monchegorsk. The area with a destroyed plant cover is coloured red in the picture. Researcher Kari Mikkola has analysed and produced the picture from a Landsat satellite image.

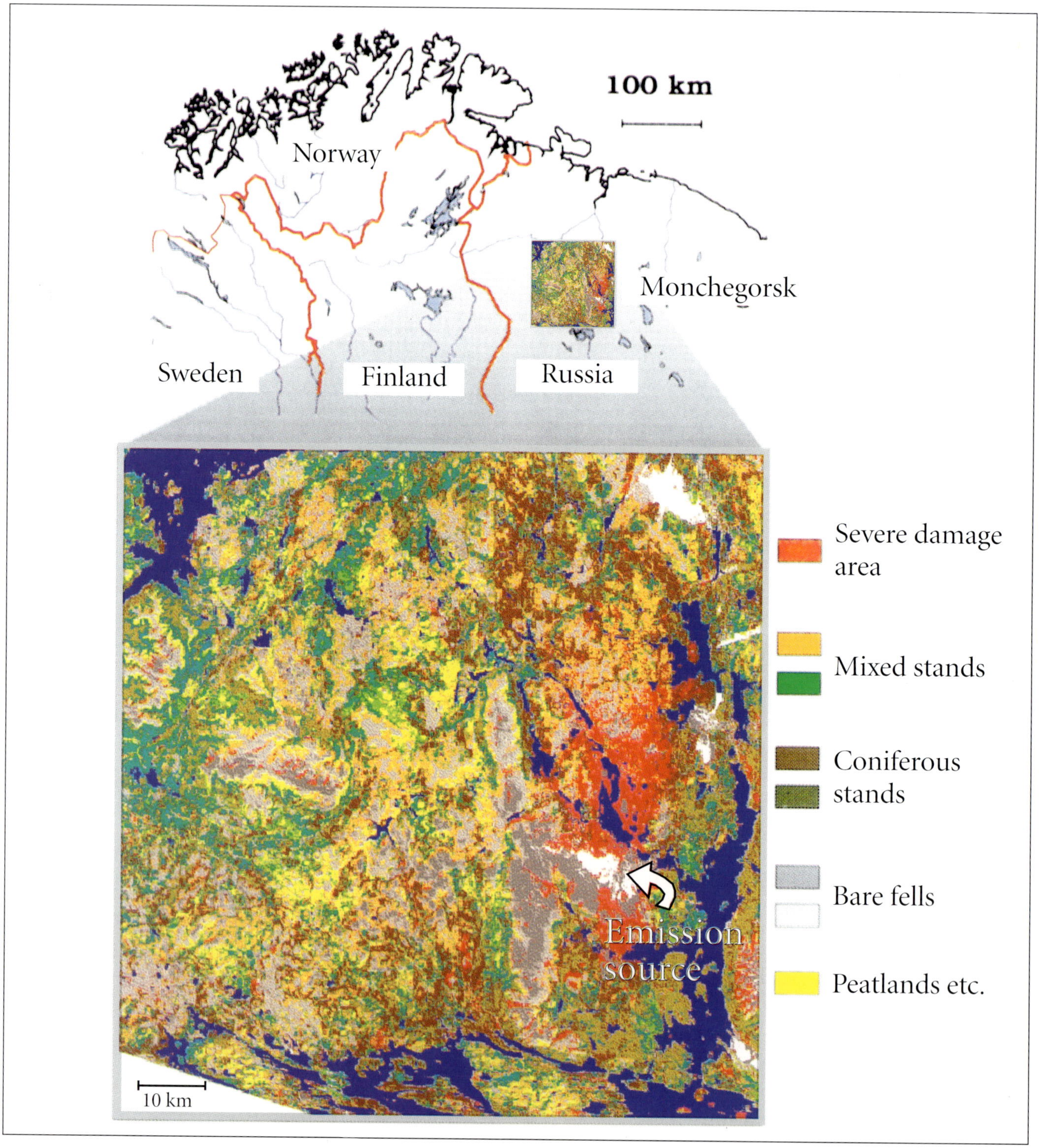

FORESTRY SUPPORTING SUSTAINABLE DEVELOPMENT

The growing world population is exploiting natural resources with methods that endanger their durability and renewability. The area of tropical forest is estimated to be decreasing at a rate of 17 million ha a year. Unless the exploitation methods are changed, the very basis of man's existence will be threatened.

A number of international agreements have already been signed concerning sustainable development. This means the adoption of industrial, economic and social developments that match the size of natural resources and take their renewability into account. Forest utilisation is sustainable when it does not exceed the biological production potential, does not prevent the continuation of ecological functioning after disturbances, nor reduce biological diversity. The prerequisites for both a diverse, rich nature and human development are also ensured for future generations.

The diverse entity comprising plant, animal and microbial species is an integral part of the functioning of the forest ecosystem, and also contributes to its maintenance. It is an intrinsic value for which man is responsible, and also a natural resource satisfying the needs of mankind.

Conserving natural resources, especially non-renewable ones, is a part of sustainable development. Pollutant emissions that damage the ecosystems have to be reduced and stopped. Net emissions of carbon dioxide have to be returned to a level corresponding to that of the natural carbon cycle. If this can be done we will be in a position to prevent climatic warming and its fatal consequences.

Man can only satisfy his needs through the use of natural resources. The increasing world population and rising standard of living in developing countries mean an inevitable increase in the production and consumption of goods.

In order to reduce carbon dioxide emissions the use of fossil fuels such as natural gas, oil and coal, and the manufacture of goods using these sources of energy, will have to be reduced. All wasteful forms of natural resource utilisation will have to be stopped.

The forest is a great, renewable and diverse natural resource. Increasing forest utilisation can therefore play a central role in sustainable development.

The production and use of wood raw materials are a part of the natural carbon and solar energy cycles. The sustained production of wood and the use of the energy it contains does not result in net emissions of carbon into the atmosphere. The carbon released from the wood is reassimilated in the growing trees. When wood replaces fossil fuels, the net carbon emissions are reduced.

Sustainable development also involves the afforestation of treeless areas and increasing the amount of tree biomass in under-stocked forest. The increase in biomass means more carbon assimilation which, in turn, reduces the net emissions. However, this is only possible up to a certain level because, as the biomass accumulates, tree mortality and the decomposition of dead biomass in the forest start to release carbon.

A natural forest not subjected to normal forest management practices is in a state of equilibrium where the amount of carbon bound in assimilation is equal to that liberated through the decomposition of dead biomass. The cycling of carbon and solar energy is a wasted from man's point of view.

The cycling of wood and wood-based products saves primary raw materials. The production of energy by burning waste wood and non-cyclable products is a part of the natural cycling of carbon.

The manufacture of durable articles from wood, such as

buildings and other structures, furniture and books, binds carbon and reduces the net emissions of carbon. The storage of garbage in anaerobic conditions has the same effect because they do not decompose. After a transitory period during which the consumption of fossil fuels decreases, the growing and utilisation of wood will be part of the overall balance between man and natural resources.

The global forest balance sheet shows that at least one half of the current growth of stemwood is left to rot in the forest without being utilised to satisfy our material needs. Millions of hectares of land are available for afforestation; in Europe alone (excluding the area of the former Soviet Union) there are at least 40 million ha if we include the agricultural land area not needed to maintain self-sufficiency in foodstuff production.

The forest is ecologically sustainable when its capacity to recover after forest fires, cuttings or other disturbances is retained. The boreal coniferous forest is one of the most durable and recoverable plant community on the globe. Apart from the areas permanently reserved for other forms of land use, forest treatment in Finland has not prevented recovery in the form of forest succession based on the laws of fire ecology. This is also the case in the northernmost parts of the country. Cuttings have not pulled back the northern timber line. It is now moving to the north and up the slopes of the fells as a result of climatic warming during the 1930's. This development has been speeded up by artificial regeneration.

Exceptions to the normal pattern of forest recovery occur, for instance, in areas where removal of the tree stand without any ditching has caused paludification of the soil. Recovery is slow or inhibited in the permafrost areas of Asiatic Russia.

There are three levels to the biodiversity of the forest:

1. Genetic variation of the trees.
2. Floral and faunistic diversity, i.e. the number of the species and their individuals.
3. The diversity of ecosystems, which consists of the habitats necessary for the species.

The treeless or almost treeless land, the forest regeneration which starts on such land, the stages of forest succession, and the mature and old, climax tree stands, form the individual components of the diversity of boreal coniferous forests. The plant and animal species have become adapted to following their own biotopes as they move with respect to space and time.

What is nowadays appreciated as diversity is primarily derived from the days of forest fires, slash-and-burn agriculture and forest grazing. Now that these diversifying factors have been removed and cannot be restored, the growing and harvesting of wood in accordance with the laws of fire ecology are the only means of maintaining overall biodiversity. The biotopes of nature conservation areas and patches of virgin forest are an essential part of this whole.

The people responsible for forestry maintain biodiversity by retaining and creating the biotopes that form it. They should know the rare, threatened species and the environments in which they live.

In 1990 there were 1 692 rare, officially threatened species in Finland. 138 species had become extinct, 217 were partially threatened, 308 threatened, and 1029 were under surveillance. 43 % of these species live in the forest. This is a relatively low figure compared to the situation in areas under other forms of land use. Forest covers 76 % of the land area.

We do not really know whether the scarcity of rare and endangered species living in the forest is due to natural dynamics of the species, pollution, illegal trapping or collecting, or forest management. The protection of the flora and fauna presupposes more research and co-operation between nature conservationists and those responsible for forestry.

When we take into account the effective control of forest fires and the changes that have occurred in land use, it is clear that biodiversity would have been smaller if forestry had not established stands on treeless or almost treeless land. The genetic diversity of the trees in stands established by sowing or planting, which always include naturally regenerated seedlings, is greater than that in stands which have developed only from natural seedling material.

The peatlands ditched for forestry have a greater biodiversity than the same peatland types in nature. The stands of exotic tree species established for research purposes also increase biodiversity.

Silvicultural methods are being developed which retain and increase the valuable components of biodiversity. At the same time as the cultural biotopes left by earlier forms of land use, such as the stands on slash-and-burn land and grazing land, become smaller in number and disappear, forest management is creating new types of cultural biotope. The surroundings of springs and streams, river and lake banks, wetlands and rare broad-leaved trees and shrubs are left in areas where forestry is not practised. About 40 % of the original area of wetlands will be left in a natural state.

Rotting trees have their own specific flora and fauna. In addition to the fact that over half of the biomass is left to rot in commercial forests in the form of stumps, roots, cutting residues, branches, needles and leaves, trees are left in the forest during timber harvesting that are worthless or comprise only low-value timber. Old, rot-ridden broadleaves, especially aspens, are important for maintaining the valuable components of biodiversity.

In Finland, as in other countries, the amount of protected forest is increasing. Efficient tree-growing in commercial forests reduces the need to start utilising the remaining natural forests. The possibility of establishing conservation forest in Finland is now much greater when the annual growth of stemwood is about 80 million m^3, than would have been the case, in the absence of silvicultural inputs, if it had remained at 50 million m^3.

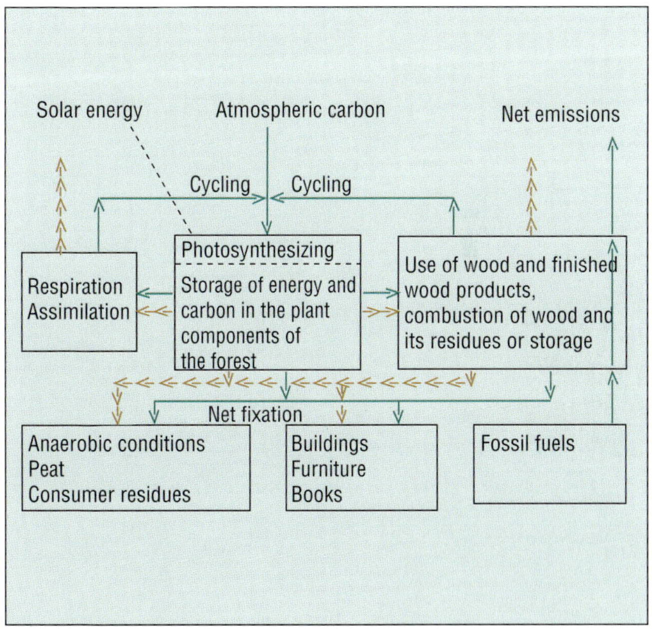

Througflow of energy (→→→) and carbon cycling between the atmosphere and the forest (→), use of wood, finished wood products, and wood residues, consumption of fossil fuels as a cause of net carbon emissions, and net carbon fixation as wood in products and dumping of consumer residues in anaerobic conditions.

This clear-cut area has been burnt and harrowed in northern Häme. The soil is glacial till and the site of the Myrtillus forest site type. Old pines and broad-leaf trees were left in the area. The wet depression containing broad-leaves and spruce was not treated. The method has created the natural pioneer forest stage which follows a forest fire.

FLORAL AND FAUNAL SUSTAINABILITY IN THE MANAGEMENT OF COMMERCIAL FORESTS

Finland has signed commitments to practice ecologically sustainable forestry that ensures the maintenance of biodiversity in the forests. The silvicultural methods and recommendations were revised by the forestry authorities and organisations at the beginning of the 1990's. One advantage for commercial forest management in the boreal coniferous forest zone compared to the other forest zones is the relative paucity of the flora and fauna. In Finland there are an estimated 45 000 species, half of which live in the forests. Depending on how they are classified, there are 300 - 850 rare or locally threatened forest species. Forest ecology is characterised by its great dynamism. The habitats move with respect to time and space as a part of the succession cycle, and the species characteristic of each habitat follow the change. Habitats which have for long been protected from fires and other foms of disturbance have their own specific species which are rare elsewhere, and are hence worth preserving.

Site classification based on forest site types is of great assistance in delineating cutting areas in the field and in estimating fire susceptibility. Owing to the fragmented land-ownership structure and topography, and small size of stand compartments in southern Finland, the clear-cut areas are usually small. The mean size of clear-cut areas is 2 ha, and in northern Finland somewhat larger. In topographically variable terrain, naturally burnt areas are usually small and fragmented.

Ecological sustainability can be estimated on the basis of a biotope's sensitivity to fire. The greater the likelihood that an area will be burnt under natural conditions, the lower is the number of species specialised to the biotope in question. On the other hand, biotopes which do not burn, or which only burn under exceptional conditions, contain highly specialised species whose survival is dependent on the stability of their habitat. Fire sensitivity provides one model for estimating the ecological sustainability of a habitat. In susceptible areas the flora and fauna are poor and rapidly recover.

The size of the managed area does not set a limit on the ecological sustainability of fire-sensitive biotopes. Fresh and dry upland sites, which account for the majority of the forest land area, are fire-prone habitats. The largest forest fires occur under natural conditions on dry upland sites covered in pine forest.

Biotopes which are valuable from the point of view of species richness are herb-rich sites, small virgin forests, old broad-leafed forest, fertile spruce mires, forest islands surrounded by swamp, small watercourses, bluffs and cultural biotopes. The special requirements of such areas are taken into account in cutting activities, for example by regulating the tree canopy coverage.

The edge zones in the terrain form refuges for many species, and attempts are made to create and develop them through cuttings. These include the banks of watercourses, roadsides, stream banks and the edges of wetlands. They are the ecocorridors which provide the species with passage from one fire refuge to another.

There are many springs along the edge of the Salpausselkä esker in southern Finland. The groves growing around these springs are protected from silvicultural treatment because they contain diverse coniferous forest ecosystems with a specialised flora and fauna.

Seed trees have been left on the top of these cliffs to promote regeneration and landscape management in Finland's lake district.

Light-demanding species benefit from cuttings. An eastern species in Finland, bluesuckle (Lonicera pallasii), grows in Kuusamo in northern Finland, and is extremely rare. It has flourished as a result of the cuttings and soil preparation carried out along the river.

Calypso (Calypso bulbosa) is the beauty of the boreal coniferous forest. It grow in old, calcium-rich, mixed pine-spruce forest. The species is fairly common on calciferous soils in northern Finland.

References:

Eronen, M. 1990. Muuttuva ilmasto (The changing climate). Terra 102:4, pp. 220-238.

Eronen, M. 1991. Jääkausien jäljillä. Tähtitieteellinen yhdistys Ursa. Julkaisu 43. Rt-Paino. Pieksämäki.

Eyre, S. R. 1968. Vegetation and soils. A World Picture. Edward Arnold. London.

Faugli, P., Lund, C. & Rye, N. 1991. Jostedalen. Natur - vannkraft. Summary: Jostedalen. Natural sciences - hydropower. Strandberg & Nilsen A.S. Oslo.

Granström, A. 1991. Elden och dess följeväxter i södra Sverige. Skog & Forskning nr 4. Sveriges Skogsförbund.

Haataja, V. 1993. Tuntsan palo ja suuri nokisavotta. Koillismaan Kirjapaino OY. Kuusamo.

Heikinheimo, O. 1915. Kaskiviljelyn vaikutus Suomen metsiin. Acta Forestalia Fennica 4.

Helander, A. Benj. 1949. Suomen metsätalouden historia. WSOY. Porvoo-Helsinki.

Ilvessalo, Y. 1920. Kasvu- ja tuottotaulukot Suomen eteläpuoliskon mänty-, kuusi- ja koivumetsille. Referat: Ertragstafeln für Kiefern-, Fichten- und Birkenbestände in der Südhälfte von Finnland. Acta Forestalia Fennica 15.

Ilvessalo, L. & Jalava, M. 1930. Maapallon metsävarat. Summary: Forest resources of the world. Communicationes ex Instituto Quaestionum Forestalium Finlandiae 16.

Johnson, E. A. 1992. Fire and vegetation dynamics. Cambridge University Press. Cambridge.

Kalela, E. K. 1961. Metsät ja metsien hoito. WSOY. Porvoo-Helsinki.

Kalliola, R. 1973. Suomen kasvimaantiede. WSOY. Porvoo-Helsinki.

Karjalainen, T. , Kellomäki, S., Lauhanen, R. & Tuovinen, J. 1991. Ilmaston muutoksen vaikutus metsäekosysteemiin ja metsänkäyttöön: Mekanismeja ja kehityssuuntia. Silva Carelica 19.

Kellomäki, S. 1991. Metsäekologia. Silva Carelica 7.

Koivisto, P. 1959 (koonnut). Kasvu- ja tuottotaulukoita. Growth and yield tables (collected by Pentti Koivisto). Metsäntutkimuslaitoksen julkaisuja 51.8.

Korhola, A. 1990. Suomen soiden synty ja kehitys (The origin and development of the mires in Finland). Terra 102:4, pp. 256-267.

Korhola, A. 1990. Suomen metsien kehitysvaiheet (The development of the Finnish forests). Terra 102:4, pp. 268-274.

Kuusela, K. 1972. Suomen metsävarat ja metsien omistus 1964-70 sekä niiden kehittyminen 1920-70. Summary: Forest resources and their ownership in Finland 1964-70 and their development 1920-70. Metsäntutkimuslaitoksen julkaisuja 76.5.

Kuusela, K. 1990. The dynamics of boreal conifer forests. Finnish National Fund for Research and Development, SITRA 112. Helsinki. (A Russian edition published by Repola, 1991). Gummerus Kirjapaino OY. Jyväskylä.

Kuusela, K. 1994. Forest resources in Europe. European Forest Institute. Research Report 1. Cambridge University Press. Biddes Ltd. Guildford, Surrey, Great Britain.

Kuusela, K. & Salminen, S. 1991. Suomen metsävarat 1977-1984 ja niiden kehittyminen 1952-1980. Summary: Forest Resources of Finland in 1977-1984 and their development in 1952-1980. Acta Forestalia Fennica 220.

Labau, V. J. & van Hees, W.S. 1990. An inventory of Alaska's boreal forests. Their extent, condition, and potential use. U.S. Department of Agriculture, Forest Service Anchorage, Alaska.

Laitakari, E. 1960. Metsähallinnon vuosisataistaival 1859-1959. Silva Fennica 107.

Larsen, J.A. 1980. The boreal ecosystem. Academic Press, Inc. New York.

Lehto, J. 1969. Käytännön metsätyypit. Kirjayhtymä. Helsinki.

Lesnaya industriya SSSR 1980. Moskva: Izdatelstvo "Lesnaya promyshlennost".

Lowe, J.J., Power, K. and Gray, S.L. 1994. Canada's Forest Inventory 1991. Canadian Forest Service.

Manker, E. 1977. Kahdeksan vuodenajan kansa. Svenska Förlags AB Nordbok.

Mayer, H. 1984. Wälder Europas. Gustav Fischer Verlag. Stuttgart.

METSÄ 2000 -ohjelman tarkistustoimikunnan mietintö, 1992. Komiteanmietintö 1992:5. Helsinki.

Nyyssönen, A. 1958. Kiertoaika ja sen määrittäminen. Summary: Rotation and its determination. Metsäntutkimuslaitoksen julkaisuja 49.

Packham, J.R. & Harding, D.J.L. 1982. Ecology of woodland processes. Edward Arno Ltd. London.

Pietikäinen, J. & Fritze, H. 1993. Mikrobial biomass and activity in the humus layer following burning: short-term effects of two different fires. Can. J. For. Res. 23:1275-1285.

Schmidt - Vogt, H. 1983. Nadelwald - ökosysteme. Allgemeine Forst Zeitschrift. Nr. 50, s. 1355-1359.

Statisticheskii sbornik 1990. Lesnoi fond SSSR. (Po uchetu na 1 yanvarya 1988 goda). Moskva: Gosudarstvennyi Komitet SSSR po lesu.

Taipale, K. & Saarnisto, M. 1991. Tulivuorista jääkausiin. Suomen maankamaran kehitys. WSOY. Porvoo.

Tikkanen, E. 1995 (toim.). Kuolan saastepäästöt Lapin metsien rasitteena, 1995. Itä-Lapin metsävaurioprojektin loppuraportti. Gummerus. Jyväskylä.

Tseplgaev, V.P. 1961. The forests of USSR. Translation. Jerusalem.

UN. Yearbook of forest products. FAO Forestry Series. Rome.

UN 1992. The Forest Resources of the Tempperate Zones, Volumes 1 and 2. UN - ECE/FAO. New York.

Vuokila, Y. & Väliaho, H. 1980. Viljeltyjen havumetsiköiden kasvatusmallit. Summary: Growth and yield models for conifer cultures in Finland. Metsäntutkimuslaitoksen julkaisuja 99.2.

Windhorst, H.-W. 1978. Geographie der Wald- und Forstwirtschaft. B.G. Teubner. Stuttgart.

Kullervo Kuusela
(born 1924), Professor Emeritus, Research Coordinator at the European Forest Research Institute. In 1962–1988 he was a professor at the Finnish Forest Research Institute responsible of the National Forest Inventories and Forest Management Planning on state and regional levels. In 1988-1993 he worked as a forest resource expert at Jaakko Pöyry Companies in forestry projects in Scandinavia and the European Russia. His latest publications comprise dynamics of boreal forests and long-term development of European forests and their utilization.

EUROPEAN FOREST INSTITUTE
Torikatu 34, FIN-80100 Joensuu, Finland
Tel. +358-73-124 395
Fax +358-73-124 393

Simo Hannelius
(born 1943) has worked for many years at the Finnish Forest Research Institute as a research specialist on the economic questions of commercial forest production. He has published a number of scientific papers and textbooks on these topics. His newspaper articles on the economical utilisation of forests and the development of forest nature, as well as courses arranged for forest owners and university students, have contributed considerably to the public's understanding of these problems. Forest photography has become almost a second profession for him.

FOREST RESEARCH INSTITUTE
Jokiniemenkuja 1, FIN-01300 Vantaa, Finland
Tel. +358-0-8570 5322, Mobile +358-49-703 147
Fax +358-0-8570 5361